Orchid

Orchid

A Cultural History

Jim Endersby

The University of Chicago Press
Chicago and London

Kew Publishing
Royal Botanic Gardens, Kew

The University of Chicago Press, Chicago 60637
The University of Chicago Press, Ltd., London
Text © 2016 by The University of Chicago
Illustrations © The Board of Trustees of the Royal
Botanic Gardens, Kew, unless otherwise noted.

Published in 2016 by The University of Chicago Press
and
Published in 2016 by the Royal Botanic Gardens, Kew,
Royal Botanic Gardens, Kew, Richmond, Surrey, TW9 3AB, UK
www.kew.org

25 24 23 22 21 20 19 18 17 16   1 2 3 4 5

The University of Chicago Press ISBN-13: 978-0-226-37632-5 (cloth)
The University of Chicago Press ISBN-13: 978-0-226-42703-4 (e-book)
Kew Publishing ISBN: 978 1 84246 629 2

Library of Congress Control Number: 2016018415
British Library Cataloguing in Publication Data
A catalogue record for this book is available from the British Library.

♾ This paper meets the requirements of ANSI/NISO Z39.48-1992
(Permanence of Paper).

Printed in the United States of America

For Pam

Contents

Introduction
Imagining Orchids

A particular flower may make us think of love or heaven, it might symbolize anything from a political cause to a drug craving, or remind us of its usefulness as a food, for medicine, or purely for decoration. Every plant that humans have ever taken a close interest in has inevitably picked up its own cultural associations, so it is not surprising that flowers as unusual as orchids have acquired their own very specific set of images, ideas, and symbols. But the particular significances we attach to orchids may be the strangest ever to have become linked to a plant.

We get a glimpse of some of the bizarre and extraordinary ways in which orchids feature in our imaginations by looking at two very different films. In *Moonraker* (1979), the eleventh James Bond movie, the evil villain Hugo Drax is planning to annihilate humanity with a deadly nerve gas that is derived from an imaginary species of South American orchid, *Orchidae Nigra*. When the intelligence service's resident scientific expert "Q" identifies the flower, Bond comments that it's "a very rare orchid indeed," thought to be extinct until a missionary found some deep in the Amazon jungle (Bond even corrects Q as to the precise location). Orchids have often been thought of as floral aristocrats, rarefied and elite, so their expert knowledge of these flowers creates a kind of kinship between Bond and Drax, marking them both as cul-

Figure 1. John Laroche (Chris Cooper) with the fabulous Ghost Orchid.
(From *Adaptation*, 2002)

tured, sophisticated men. When Bond arrives in the villain's traditional, lavishly appointed underground hideout, Drax naturally starts pontificating about his secret plan for world domination (luckily for our hero, none of his opponents ever thinks to just shoot him — especially inexplicable when Roger Moore was playing the role). Drax points pompously to the orchids that hang above him: "the curse of a civilisation," he explains. The indigenous people who built the ruined city within which his lair is hidden worshipped the flower, but long-term exposure to its pollen caused sterility. Drax boasts that he has "improved" upon sterility; "those same seeds now yield death," so his plan to gas the world and then repopulate with his own hand-picked master race is called "Operation Orchid."

The idea of lethal orchids in the jungle had been around for centuries before Bond's screenwriters used it. Indeed, lethal orchids were, as we shall see, one of many exotic components that made up the imaginary place that Europeans came to call "jungle." The word itself derives from Sanskrit, via the Hindi word *jangal*, which really means barren waste-ground, including deserts, but over many centuries the idea of the "jungle" became the centerpiece of European fantasies about the tropics: steamy and dangerous, but filled with elusive treasures such as orchids, a place far from civilization where white men could go to prove they were men — or die in the attempt.

More than twenty years after *Moonraker*, the film *Adaptation* (2002) used orchids in very different ways, yet there were some surprising similarities. *Adaptation* was based on Susan Orlean's best-selling book *The Orchid Thief*, which centered on John Laroche, a real-life, modern orchid thief whose arrest for stealing endangered Ghost Orchids (*Dendrophylax lindenii*) from a nature reserve initially sparked Orlean's interest in these flowers. As she researched and wrote the book, Orlean came to see Laroche's hunt for the flower as a symbol of obsessive desire, a desire whose real goal seemed not to be the orchid, but simply to possess something almost unobtainable. As she wrote, "I suppose I do have one embarrassing passion — I want to know what it feels like to care about something passionately."[1] The movie is built around the problem of filming an unfilmable book, a problem it solves by creating a series of witty analogies for Orlean's "embarrassing passion." The director and writers create a fictitious version of Orlean (played by Meryl Streep) and then imagine this handsome, sophisticated woman from the big city becoming romantically obsessed with the strikingly unkempt and earthy backwoodsman, Laroche (played by Chris Cooper). The film sets up a striking visual contrast between the flower's delicate beauty and Laroche's snaggle-toothed toughness, his manly willingness to wade into seemingly impassable swamps (an updated jungle fantasy) in order to outwit the law. The fragile (and ultimately unobtainable) orchid, an evanescent ghost so rare and close to extinction that it seems almost imaginary, contrasts strikingly with Laroche himself, who is portrayed as physically rather repulsive. Yet his desire for the orchid makes him curiously desirable; the contrast between his lousy teeth and the exquisite pale flower mirrors the perverse mixture of attraction and repulsion that orchids themselves seem to provoke in some people.

These two films provide a glimpse of the recent meanings of orchids. They are deadly but endangered. Alluring, even sexy. Luxurious and expensive. Delicate "hothouse" flowers, not really adapted to the real world, yet they can be lethal. Mostly found in mysterious distant jungles, orchids have often been depicted as feminine and delicate, which may be why larger-than-life male heroes have to be imagined as the only ones who can bring them home safely. People will commit crimes for them, even kill or die for orchids.

It has taken over two thousand years for orchids to become the reflection and object of such a diverse range of human desires. For the past couple of hundred years, they have usually been thought of as effete exotics, pointless luxuries that aptly symbolize the idle rich — on whom they depend for their hothouses — whereas in the twenty-first century they are more likely to be seen as endangered

Figure 2. A real Ghost Orchid, *Dendrophylax lindenii*, drawn by Blanche Ames. (From Oakes Ames, *Orchids in Retrospect*, 1948)

rarities, emblematic of the urgent need to conserve fragile ecosystems. Yet these different images of the orchid share common roots: the rarity of many orchids helped make them into expensive trophies for wealthy collectors, while the imperial networks that brought them to European hothouses would eventually contribute to the destruction of their once-inaccessible habitats.

This book will follow orchids through history, commerce, art, literature, science, and cinema, from Greece around 300 BCE to the chalk downland on Britain's south coast today, in order to discover when, why, and how they came to have their cultural associations. In the process, we will discover some unexpected connections; everything from the European conquest of the Americas to Charles Darwin's theory of evolution proves to be intimately ensnared with the story of the orchid.

The heart of this book is the history of our scientific understanding of orchids. It took centuries for science simply to realize that orchids were indeed a single family of plants, and many mysteries about their biology remained unsolved until quite recently. So, if you're not quite sure what an orchid is, how science defines the group, don't worry—you are in good company because for most of our story, nobody knew exactly what orchids were. Over successive chapters, we will discover them in much the same way as history's scientific investigators did. (However, if you want to jump ahead and get an overview of orchid biology and anatomy, you'll find the essentials in the section "Every Trifling Detail," chapter 5.) As Europeans traveled, traded, and conquered ever more widely, they gave names to all the new places, peoples, animals, and plants they were discovering, but this eruption of new things with new names rapidly became confusing. The same plant often had many different names in different places, which became a pressing concern when the flower was potentially valuable.

The modern scientific names of orchids (and all other living things) owe their origin to the eighteenth-century Swedish naturalist Carl Linnaeus, who synthesized the diverse achievements of his predecessors into a unified, simple, and robust system for naming the newly enlarged world's profusion of living things. Somewhat surprisingly to modern minds, the sexuality of plants was at the heart of Linnaeus' original system for naming them; he saw the plants in very human terms, as "husbands" and "wives" united by the "marriages of flowers." (At times, Linnaeus even struggled to find his own, strict moral code embodied in the flowerbeds.) In the following century, the sex lives of plants were one of the factors that brought them to Darwin's attention; he spent many years in his garden and greenhouse, puzzling over why orchids go to such elaborate lengths to build an intimate relationship with the particular species of insect that pollinate them. Darwin's research was helped by the prevailing orchidmania; the high prices paid for rare orchids allowed a whole industry of collectors, shippers, nurserymen, writers, and illustrators to grow up around the flowers. The imperial trading networks, protected by Britain's navy, that brought new orchids to Britain allowed Darwin to extend his research from native British species to exotic tropical ones. And when he wrote his book about them, the Victorian fashion for orchids ensured that his ideas reached a wide readership (even though his book was a rather dry and technical botanical treatise). Yet some of Darwin's readers were taken aback by the book's unexpected religious message. For many centuries, the fact that flowers such as orchids relied on insects to pollinate them and supplied the insects with nectar in exchange, had been used to argue not only that God must exist (to design such perfect creatures), but also that He was clearly benevolent, caring for orchids and insects yet adding a gratuitous garnish of beauty for humans to enjoy. Darwin's work is usually thought of as contributing to the destruction of this comforting picture and some of his readers were indeed distressed by the picture he painted, but others drew unexpected consolation from Darwin's orchids precisely because God had *not* made them.

The many ways humans have used, grown, collected, and studied orchids have shaped—and have been shaped by—the ways in which we have imagined orchids. Orchid flowers are extraordinarily diverse in their sizes, shapes, and colors. Some are elegant, others grotesque (and no two orchid lovers would agree on which are which). Some seem to mimic insects, animals, or human body parts (the significance of such mimicry was a mystery even Darwin couldn't solve; it took three twentieth-century naturalists—a French judge, an English colonel, and an Australian schoolteacher—to discover what Darwin had missed). Partly because orchids take such extraordinary shapes, they have cropped up repeatedly in the stories we humans tell ourselves. Whether it's in films, novels, plays, or poems, orchids feature repeatedly from Shakespeare to science fiction, from hard-boiled thrillers to elaborate modernist novels, and tracking orchids across diverse genres reveals unexpected connections. For example, orchids crop up in the late nineteenth century in tales of imperial derring-do; a hunt for orchids often provided the pretext for adventurous heroes to face death in the pages of the *Boy's Own Paper* and the novels of H. Rider Haggard (who wrote Edwar-

dian bestsellers including *She* and *King's Solomon's Mines*). Unexplored tropical jungles had provided writers like Haggard with infinite scope, a dark can-
vas on which to depict imaginary tribes of cannibals, scheming and seductive
native women, and other appealing mysteries. But, writing at the end of Queen
Victoria's reign, Haggard realized that the blank spaces on the map were being
filled up; where, he wondered, would future writers turn for appropriately adventurous and romantic settings? His contemporary, the arch-imperialist Cecil
Rhodes, inadvertently provided the answer when he told a journalist that since
the Earth is "nearly all parcelled out," he would annex the distant planets if he
could.[2] And, sure enough, in the early twentieth century, science fiction took
orchids into space, where they were used to devise even more titillating and unexpected adventures.

As orchids were used and reused in different fictions across many centuries,
they were often linked to sex, symbolizing romantic intentions, or seductive de-
signs afoot. They were once believed to have aphrodisiac properties (although,
rather unexpectedly, they were thought to both provoke *and* curtail lust), and
some of these myths seem to have traveled with Europeans when they discovered the New World. Among the rare and curious novelties of America was
vanilla, which is made from the cured seed pods of an orchid and, until very
recently (when science allowed the mass production of "airport orchids"), was
the only orchid with any real commercial value. Vanilla—like many New World
products—was believed to be an aphrodisiac, largely because of its associations
with the sensuous tropics. Ironically, vanilla wasn't immediately recognized as
an orchid, so perhaps its imaginary aphrodisiac properties influenced those who
eventually classified it. As we survey some of the many ways orchids have been
imbued with erotic connotations, it becomes apparent that sexy orchids often
tell us more about ourselves than they do about the sex lives of plants. Writers
and filmmakers have used orchids to symbolize everything from conventional
marriage to homosexuality; they have been deployed to suggest everything from
male discomfort when faced with sexually assertive women, to a celebration of
sadomasochism.

This book will explore some of the curious and unexpected variety of significances that people have ascribed to orchids by focusing on European cultures
(and those that Europeans have decisively shaped), where two clear themes
emerged that link many of the orchids' cultural associations—sex and death—
ideas that recur in the West's imaginings of orchids. Ironically, the association

with sex began with a coincidence; most Mediterranean orchid species have the paired tubers that originally gave them their name (the Greek word *orkhis* means "testicle") because they had to endure semi-drought each summer; the tubers stored food that allowed the orchids to survive until the rain returned. The majority of the world's orchids grow in the tropics and seldom face this problem, and hence most don't have tubers. If the story of the orchid had begun anywhere else, the link with human sexuality might never have been made (as, indeed, it never seems to have been in traditional Chinese and Japanese orchid culture, where orchids have been grown and tended for many centuries). The orchid's sexy, deadly reputation is in turn linked to the fact that Europeans have (for better or, more frequently, for worse) been the world's most active travelers, conquerors, and colonizers; orchids are a good example of the exotic riches that Europeans sought—and would kill for—and which shaped their vision of the new worlds they set out to annex. Although science is a global enterprise today, it remains dominated by a tradition that started on the shores of the Mediterranean, where the orchid first received its name. And it is, of course, no accident that imperialism and Western science came into the world at around the same time and place; they were born joined at the hip, and have never really been separated. But that's a topic for another book.

People love orchids for their beauty, of course, the enthralling variety of their colors and shapes are the key reasons we grow them, but there is much more to orchids than meets the human eye. Science, empire, sex, and death have shaped the ways we understand orchids; as our cultures have changed, so too have the ways we see orchids. And in turn the orchids have shaped human cultures, including our sciences; the full intimacy of the relationships between orchids and insects was only realized in the twentieth century when plant biologists discovered that orchids cheat their insects, seducing them into pollination without rewarding them with nectar. The orchids' extraordinary tactics were not recognized until the plants had been reimagined as active, cunning, and seductive—a process that owed as much to the father of modern science fiction, H. G. Wells, as it did to Darwin. Thanks to the unexpected connections between science and fiction, the orchids' tricks for propagating themselves were discovered, which now allows ecologists to use the "lock and key" fit between particular orchids and their pollinating insects as an acutely sensitive measure of the impact of climate change.

Plants like orchids are usually considered part of nature — a world that exists outside and independently of us — and which we usually contrast with culture — the world humans have created. Yet there is no stable boundary between the natural and the cultural; we cross, erase, and redraw that frontier whenever we imagine orchids.

1

Censored Origins

The Loeb Classical Library publishes editions of the Greek and Roman classics that have shaped so much of Western culture. Each of their books includes the original language alongside an English translation and—because they are widely used in schools and universities—their early editors faced a rather delicate problem; some of these classics contained passages that were deemed (at least in the early twentieth century when the series began) to be unsuitable for impressionable young minds. Take the Athenian comedian Aristophanes, for example. His celebrated play *Lysistrata* (or the "army-disbander") is about the women of Athens and Sparta trying to put an end to the endless wars between their cities. They hit on the simple but effective strategy of denying their men sex until they stop fighting. This leads to a wonderfully bawdy scene when the men finally meet to negotiate; each is unable to assuage the permanent erection he has acquired as a result of his enforced abstinence. The text is full of lewd puns and double-entendres whose meanings would have been made unmistakable by the Ancient Greek actors' costumes, which frequently featured gigantic fake penises. What was a respectable publishing house to do with such material?

The publisher's solution for material like that in *Lysistrata* was simply to leave the offending passages in the original Latin or Greek (thus providing generations of young students with an additional incentive to master these ancient languages). However, concealed beneath the modest green and red covers of the Loeb library there is one exception to this rule. Amid the numerous tales of rape, incest, orgies, and debauchery, only one passage was considered so unsuitable as to be omitted altogether. When Sir Arthur Hort edited the Loeb edition of Theophrastus's *Historia Plantarum* (*Enquiry into Plants*), he dropped the passage "On Aphrodisiacs and Sexual Drugs" (IX.18.3–IX.18.11); not only was there no translation—even the Greek was excised. This appears to be the only such omission in the more than 500 volumes of Loeb and the decision must have been taken quite late in the publication process, because the book retains an enigmatic index entry to one of the missing aphrodisiac plants, an orchid.[1]

To understand the sexy orchid's place in Western culture, we have to begin (as we do with so many other aspects of our science, culture, medicine, and mythology) in Ancient Greece. The Western traditions of understanding living things originated on the shores of the Mediterranean; for animals, we turn to Aristotle, but for plants, we turn to his friend and successor, Theophrastus of Eresus (ca. 371–ca. 287 BCE), universally acknowledged as the "father of botany." The title sounds like quite an honor, especially if you are remembered more than 2,000 years after your death, yet Theophrastus might well be disappointed if he were able to know of his fame. Although he clearly loved and was fascinated by plants, for him they were merely one among many studies. He was a pupil of both Plato and Aristotle (who described him as "a man of extraordinary acuteness, who could both comprehend and explain everything").[2]

The ninth book of Theophrastus' *Enquiry into Plants* is a surprising text to be the target of censorship, since it's the world's earliest surviving Greek herbal and provides a unique insight into the Ancient Greeks' knowledge of plants and their uses. In amongst many other topics, Theophrastus comments that, "Besides affecting health, disease and death, people say that some plants have other specific effects, physical and mental. Firstly, the physical effects, by which I mean favouring or inhibiting procreation":

> There is at least one plant whose root is said to show both powers. This is the so-called salep, which has a double bulb, one large and one small. The larger, given in the milk of a mountain goat, produces more vigour in sexual intercourse; the smaller inhibits and forestalls.[3]

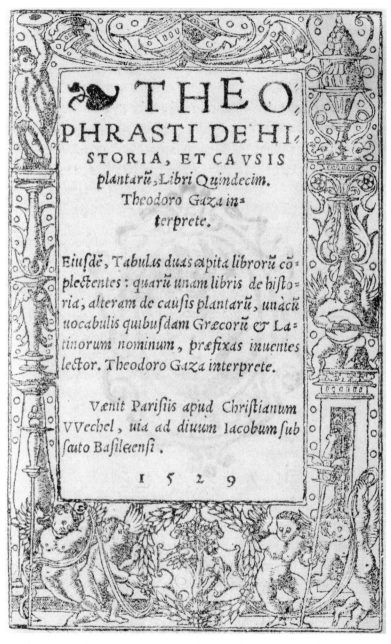

THEO
PHRASTI DE HI-
STORIA, ET CAVSIS
plantarū, Libri Quindecim.
Theodoro Gaza in-
terprete.

Eiufdē, Tabulas duas œpita librorū cō-
plectentes: quarū unam libris de hiſto-
ria, alteram de cauſis plantarū, unàcū
uocabulis quibuſdam Græcorū & La-
tinorum nominum, præfixas inuenies
lector. Theodoro Gaza interprete.

Vænit Pariſiis apud Chriſtianum
VVechel, uia ad diuum Iacobum ſub
ſato Baſileænſi.

1 5 2 9

Figure 3. Title page from a sixteenth-century Latin edition of Theophrastus'
botanical works (Basel, 1529). This edition was donated to Kew by Sir Arthur Hort,
the man who censored the details of orchid's aphrodisiac properties.

In this short passage, the orchid made its debut in Western science; "salep" is a kind of nutritious porridge, still widely consumed in parts of the Middle East, that is made from what are often called the bulbs of orchids (they are really tubers or rhizomes), such as the Early Purple orchid (*Orchis mascula*). The tubers contain stored starch (the larger and smoother tuber is formed by the current year's leaves; the other is last year's, becoming smaller and wrinkled as its food supply is used).[4] Thanks to Theophrastus' decision to record this folk myth about the plant's aphrodisiac properties, orchids were associated with sex from the first moment they appeared in Western writing and the link has never been broken.

Theophrastus did not coin the name orchis (from which the modern name, orchid, is derived); it—along with the information about its aphrodisiac properties—came from the root-cutters, or "rhizotomai," a group of now-nameless herbalists who were experts in the culinary and medicinal uses of roots (in modern usage, a rhizome is an underground plant stem, but the word has often been used informally to mean a root). The Ancient Greeks knew of a handful of species of orchids, all of which grew around the Mediterranean. As we have seen, in order to survive the semi-drought of summer these species had underground storage organs, which botanists describe as paired "globose tubers," but which everyone else, from the ancient world onward, recognized as resembling a pair of testicles. (An early name for one of these plants was *Cynosorchis*, which translates, rather directly, as "dog's bollocks.")

The Lesbian Boy

Theophrastus was born in Eresus on the island of Lesbos and is therefore sometimes known—rather confusingly to modern readers—as the Lesbian Boy. In the ancient world, Lesbos was famous for "the excellent quality of its products, material, artistic, and intellectual"; for generations, Greeks and Romans who wanted to bestow the highest possible praise on anything, "whether it were a piece of music, a verse of poetry, or a cask of wine, were accustomed to pronounce it Lesbian."[5] His original name was Tyrtamus and at a very young age he went to Athens to study at Plato's Academy. Tyrtamus was only 21 when Plato died and Aristotle (Plato's former pupil) inherited the Academy; it was Aristotle who renamed his friend, and former fellow-pupil, Theophrastus for the "divine character of his eloquence."[6]

THEOPHRASTI ERESII

Figure 4. An imaginary portrait of the father of botany, Theophrastus of Eresus.
(From a Dutch edition of the *Historia Plantarum*, 1644)

According to his first biographer, Theophrastus wrote more than 220 books, containing over two million words.[7] They ranged across the whole of the wide and unpredictable terrain of Ancient Greek philosophy. Among Theophrastus' interests was the nature of knowledge itself, the branch of philosophy called epistemology; he wrote a book that asked "What are the Different Manners of Acquiring Knowledge" (as well as "three on Telling Lies"). In addition to what we would still consider basic philosophy, Theophrastus wrote books on Epilepsy, Enthusiasm, and Empedocles. Not to mention "three books of Objections" (a branch of philosophy now mainly practiced by nine-year-olds). His eclectic oeuvre included a book "on Mistaken Pleasures" and one (presumably the antidote to the first) "on Pleasure according to the Definition of Aristotle."

However, for a Greek philosopher, knowledge (a word we get from the Greek *gno*, γνω) could not be contained by the kinds of modern boundaries that are created and policed by disciplinary specialists and their categories. Unhindered by any consideration of which university faculty he might belong to, or in which peer-refereed journal he might publish, Theophrastus wrote a book "on Smells" and well as "one on Wine and Oil," to say nothing of "one on Hair; one on Tyranny; three on Water; one on Sleep and Dreams; three on Friendship; two on Liberality" and "three on Nature."[8] And when it came to the branches of philosophy that we would now call *biology* (a term that would not be coined for another 2,000 years), the Lesbian boy was equally prolific, contributing a book "on the Difference of the Voices of Similar Animals; one on Sudden Appearances; one on Animals which Bite or Sting; one on such Animals as are said to be Jealous" and—just to make sure his work was really comprehensive—"seven on Animals in General."

Despite this extraordinary output, only a tiny percentage of Theophrastus' work survives, mainly fragments of his botanical works, yet they are more than sufficient to secure him the title of the world's first botanist. Two thousand years before the terms were coined, Theophrastus identified the most profound division within the plant world, that between dicotyledons (or dicots, plants with two embryonic leaves enfolded in their seeds) and monocotyledons (or monocots, those with a single seed leaf—the group that includes orchids). He described the anatomy of plants in great detail, as well as the geographical distribution of many species, he specified their medicinal and culinary properties, including their magical and aphrodisiac properties, and he even explained how to cultivate them. His knowledge was based on careful studies conducted in the garden attached to the Athenian academy (he owned at least ten slaves who gardened for him; in his will he gave some their freedom and specified that the others were to be freed when they had worked long enough).

To understand why Theophrastus deserves his place in history, consider the titles of his botanical books: the *Enquiry into Plants* and the *Causes of Plants*. The idea that plants and animals had *causes* was a new one. As far as we can tell from the very fragmentary evidence, the writers of earlier plant treatises were interested only in the uses of plants; their works were agricultural, horticultural, or medical. Although Aristotle and Theophrastus continued this tradition, they also studied living things for their own sake and they were sometimes mocked for "going about the country picking up and curiously peering into the least little things of nature, such as were of no possible use." Near the start of Theophrastus' *Enquiry* he lists the external and obvious organs of a plant such as a tree: root, stem, branch, bud, leaf, flower, and fruit. There is a logic here that is driven

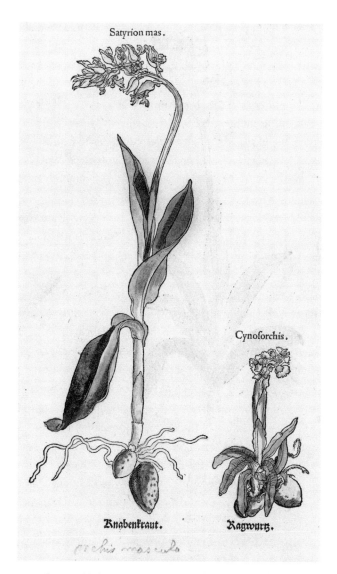

Figure 5. A botanical babel: a sixteenth-century illustration of what is now called *Orchis mascula,* showing some of its other names, including Knabenkraut, or "lad's weed." (From Otto Brunfels, *Herbarum vivæ eicones ad nature imitationem,* ca. 1530)

by curiosity, a sense that the plant no longer is a means to an end, but has become an object of interest in its own right.[9]

Although the origins of modern botany are discernible in the ancient world, we have to be careful not to separate out the knowledge that we now take seriously from all the other ways the Greeks understood plants. It is clear that Theophrastus was building on the work of earlier medicinal writers and that he shared their interest in the practical benefits people could derive from plants. The fact that his list of the plant's parts begins with roots is evidence of his debt to the earlier rhizotomai, who were obsessed with the secret, subterranean powers of roots; the mandrake that could drive you mad, and the orchid's tubers that could promote or suppress lust. Theophrastus built on this complex mixture of folk myth and practical wisdom — often accepting, occasionally questioning — but he added a rationale for the structure of his list, which was simply that "in all plants the growth of the root precedes that of the superior parts" (*Enquiry*, Book I, chapter 2). The story of the orchid begins with Theophrastus' account of its parts and their uses, but the orchid's story is part of a much wider one — that of the long, slow death of Greek philosophy. Over the next two millennia, one subject after another would be detached from the original body of knowledge; today philosophers are left to dream of far fewer things in heaven and earth than their Greek predecessors. Philosophers like Aristotle and Theophrastus took the world and everything in it, on it, below it, and above it as their object of study, from the nature of the heavens to the nature of friendship, from the birth of the cosmos to that of an insect. Their successors were to spend centuries retreating into narrower and narrower provinces of knowledge, each with its own narrowly defined set of answers to the question "What are the Different Manners of Acquiring Knowledge." But it was from among these increasingly parochial enterprises that today's sciences emerged, giving us the power to answer questions that Theophrastus and Aristotle could barely have dreamt of asking.

The Uses of Orchids

When, in the ninth book of his *Enquiry*, Theophrastus turned to the uses of plants as *pharmaka* (medicines), he recorded the apparently contradictory claim that orchids both promoted erections and caused impotence (given the orchids' similarity to testicles, it is unsurprising that female desire was not considered, yet this was an oversight male inquirers were to persevere with for many centuries). He noted that it seems surprising that a single plant should have both properties, but was convinced that "it is not absurd that there should be such powers." As evidence, he described a noted drug dealer who sold another plant that caused complete impotence: "The impotency from it can be either total or temporally delimited as, for example, two months or three months, so that it can be used on servants when one wants to punish and discipline someone."[10]

Apart from its possibly inflammatory impact on youthful readers, this kind of material may also help us understand why this passage was dropped from Hort's edition of the book. In the early twentieth century historians tended to try to squeeze Theophrastus into a history of science seen as a steady, rational march leading onward and upward to present-day truths. As one of the fathers of Western science, the great Greek had to be an epitome of reason not a peddler of folklore. (And perhaps erotic folklore was deemed particularly out of place.) So Hort may have persuaded himself that the raunchy orchid passage was of dubious authenticity, or simply in poor taste; whatever the reason, it was excised.

Yet there's no doubt that, whether or not twentieth-century historians approved, much of what is found in Theophrastus and the other great philosophers is what we would now describe as folklore. As the orchid passage makes clear, some of his information was derived from the root cutters and drug sellers and he clearly adopts their idea that the appearance of a plant might give a clue as to its uses; a plant that looks like male testicles ought to have sexual effects of some kind. (The idea that a plant's appearance gave a clue to its use would become known as the doctrine of signatures and would continue to be taken seriously by some writers for more than 1,500 years after Theophrastus' death, as we shall see in the next chapter.) Theophrastus was openly critical of many of these claims. He acknowledged that some plants are dangerous and needed to be handled carefully, but:

> The following ideas may be considered far-fetched and irrelevant; for example they say that the peony . . . should be dug up at night, for if a man does it in the day-time and is observed by a wood-pecker while he is gathering the fruit, he risks the loss of his eyesight (*Enquiry into Plants*, Book 9, chapter 8).

Like much of Greek science, Theophrastus' writing is a mixture of traditional knowledge, occasionally tempered by skepticism and by the new observations that he made in his own garden.[11]

Despite their quality, Theophrastus' writings do not appear to have been widely read in the centuries after his death, but his ideas—and the myths about the power of orchids—were perpetuated by several other writers, particularly the Roman, Gaius Plinius Secundus, better known as Pliny the Elder, whose many books included a 37-volume treatise simply called *Natural History* (*Historia Naturalis*). Writing in the first century of the common era, Pliny produced an exhaustive compilation of everything he knew about the natural world. He claimed that it contained 20,000 important facts derived from over 2,000 different books, an assertion that some modern scholars regard as an understatement.

Pliny was a noted equestrian, a cavalry general, and a busy lawyer, so busy that while eating dinner or being carried about Rome in his sedan chair, he would invariably have a servant reading aloud to him. Even in his swaying sedan chair in the middle of the crowded Roman streets, Pliny would edit the texts he was listening to, translating from Greek to Latin on the fly when necessary, and have a second servant transcribe them, ready for insertion into the next volume of the *Natural History*.[12] Unsurprisingly, there are a few errors in the book, but it is hard not to be impressed by its author's work ethic (apparently Pliny once rebuked his nephew for wasting time walking when, if he'd also been carried, he could have been reading or dictating).[13] Pliny's lack of exercise finally caught up with him; he apparently died of a heart attack after trying to rescue some of his countrymen from the eruption of Vesuvius that destroyed Pompeii. However, his books survived to become one of the richest sources of ancient knowledge and beliefs about plants, rocks, and animals. Eighteen centuries after its author's death, the French botanist Michel Adanson commented that Pliny, "the indefatigable compiler," had recorded so much classical plant lore, "in such flowery language, that one may well say of the whole that it is in beautiful disorder."[14]

Pliny clearly knew Theophrastus' works and repeated the claim that the orchid's larger or harder tuber "taken in water, is provocative of lust; while the smaller, or, in other words, the softer one, taken in goat's milk, acts as an antaphrodisiac." The particular species Pliny was describing appears to have been *Orchis italica* or *Orchis morio* (it is usually difficult to be certain with ancient botanical texts), but he also describes another species known as Satyrion (possibly *Orchis pyramidalis*) that had "the root double; the lower part, which is also the larger, promoting the conception of male issue, the upper or smaller part, that of female" (like the aphrodisiac properties, this ancient belief would persist for many centuries).[15] The confusion as to which species is intended is compounded by that fact that, as Pliny notes, "Satyrion" seems to have been a generic Greek name for any plant thought to possess aphrodisiac powers (the name comes from that of the satyrs, who in Greek mythology were bestial men, drunken and lustful, who were often the companions of Dionysus, the god of wine).

Pliny also mentions the *Cynosorchis* (dog's testicles), which has similar properties, but he adds a detail to the information about the two bulbs having opposite effects on the sex drive, which is that the practice of adding it to wine occurs "in Thessaly."[16] The mention of Thessaly (a region of northern Greece) is interesting, because Pliny's near contemporary, the Roman doctor Dioscorides, also mentions it when describing an orchid:

> The root is bulbous, somewhat long, narrow like the olive, double, one part above, the other beneath, one full but the other soft and full of wrinkles. The root is eaten (boiled) like bulbus. It is said that if the bigger root is eaten by men, it makes their offspring males, and the lesser eaten by women makes them conceive females. It is further related that women in Thessalia [Thessaly] give it to drink with goat's milk. The tenderer root is given to encourage venereal diseases, and the dry root to suppress and dissolve venereal diseases.[17]

There is no evidence that Dioscorides knew Theophrastus' work, nor were he and Pliny apparently aware of each other (although they clearly drew on common sources), so it would seem that stories about the erotic power of the orchid circulated widely in the ancient world. The women of Thessaly were supposed to be famous witches, which probably explains the idea that they could use orchids to both cause and cure venereal diseases—although this information does not occur in every version of his book, *De Materia Medica* (On Medical Materials).

Dioscorides added a couple of other kinds of orchid to those listed by his contemporaries. As usual, modern names are hard to assign, but there's one he called "serapias" (of which he noted "about this orchid there are as many stories told as of the former") and the information offered about the flowers is generally similar to that found in Pliny (Dioscorides noted, of a type he calls *Erythraïan*, that "the root is said to be aphrodisiac even when held in the hand but more so when drunk with wine.")[18] It is not surprising that Dioscorides was able

Figure 6. A Greek edition of Dioscorides, *De Materia Medica,*
printed in Venice by Aldus Manutius (1518).

to add more stories about orchids, since he traveled widely throughout the ancient Near East, having been the doctor to a Roman army legion for part of his career, and he clearly gathered information about plants wherever he went, frequently giving their names in both Greek and their original languages, such as Egyptian or Persian.

The writings of Theophrastus, Dioscorides, and Pliny were to remain the basis of Western orchid knowledge for over a thousand years. Their supposed medicinal and aphrodisiac properties probably help explain their occasional appearances in Roman architecture. For example, orchid details have been identified on the ceiling of the ruined Temple of Venus Genetrix in Rome's Forum of Caesar. Venus, in this form, was associated with fertility and motherhood, so the orchid's presence is hardly surprising, and in the few other cases where orchids appear in classical art and architecture, they seem usually to be associated with spring and fertility.[19] However, orchids do not feature in any authentic classical myths or legends — despite the numerous modern sources that claim they do.

As with most of the sciences whose origins we can trace to Ancient Greece, skepticism and the demand for first-hand evidence would eventually come to dominate botany, but when it came to orchids — as we shall see — myths, folk-lore, and superstition would persist a lot longer than with other plants. Indeed, orchids were to continue accumulating myths for two thousand years after Theophrastus first wrote about them.

Plate 1. Hugo Drax in his lair explaining "Operation Orchid" to Bond. (From *Moonraker*, 1979)

Plate 2. One of the stars of James Bateman's monstrous orchid book, *Stanhopea tigrina*, illustrated by Augusta Withers. (From *The Orchidaceæ of Mexico & Guatemala*, 1837–43)

CORYANTHES SPECIOSA. VAR.

Plate 3. The Brazilian bat orchid, *Coryanthes speciosa*, that traps bees
in its bucket-like labellum until they have pollinated it, illustrated by Sarah Drake.
(From James Bateman, *The Orchidaceæ of Mexico & Guatemala*, 1837–43)

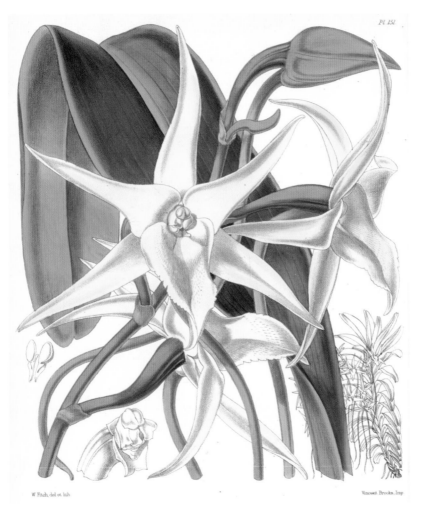

Plate 4. Probably the first specimen of *Angraecum sesquipedale* to bloom in Britain, drawn by Walter Hood Fitch. (From William Jackson Hooker, *A Century of Orchidaceous Plants Selected from Curtis's Botanical Magazine*, 1849)

Cattleya labiata

Plate 5. A fine double-page illustration of *C. labiata*; one of only two depictions of
the plant known before its rediscovery in the 1880s, illustrated by S. Holden.
(From *Paxton's Botanical Magazine*, 1837)

Red Book, Black Flower

For almost four hundred years, the Vatican library held a small book, 150 × 200 mm (6 × 8 inches) and 19 mm (3/4 inch) thick. Bound in red velvet, with gilt edged pages, it once had two metal clasps whose imprint remained clearly visible on the cover. It was written in 1552 and arrived in Rome shortly afterwards. Judging by the beautiful condition it was in when rediscovered in 1929, it was hardly used for many centuries. It is known by many names but the most common is the Badianus Manuscript, and it has two claims to major historical importance. Firstly, it is the oldest surviving record of Aztec plant lore, explaining how the indigenous people of Mexico used plants before the Spanish conquest. And secondly, it contains the first record of an orchid from the Americas, a plant from which a remedy called "The traveller's safeguard" was made.[1] In the Náhuatl language (which the Aztecs or *Mexicah*, and many other central American peoples, spoke) the plant was called *tlilxóchitl*, which was often mistranslated as "black flower." Today, we call it *Vanilla planifolia*, the orchid from which vanilla essence is derived.

Vanilla symbolizes a revolution not merely in our understanding of orchids, but of the West's understanding of nature itself. For 1,500 years the properties of orchids had remained pretty much as they had been described by Theophrastus, Pliny, and Dioscorides, but the discovery of America (in reality a rediscovery, since Viking navigators had visited much earlier) shocked many European scholars. For centuries they had been busy annotating and commenting on the ancient writers, but as the plants of the New World became known, it gradually became clear that they could not be named by reference to the classical authors. The recognition that the New World contained new plants—such as vanilla—marked the beginning of the end of hundreds of years of relative stasis in natural history.

To see how little had changed since the ancient authors wrote, consider a book published in 1568 by William Turner, physician to the Duke of Somerset. This was the first printed herbal (book of medicinal plants) in English, to which Turner gave the snappy title *A New Herball: Wherein Are Conteyned the Names of Herbes in Greke, Latin, Englysh, Duch, Frenche, and in the Potecaries* [apothecaries] *and Herbaries Latin, with the Properties Degrees and Naturall Places of the Same, Gathered and Made.* In our more impatient times, it is known simply as Turner's *New Herball*, but the adjective "new" is a little misleading. If you read it, you will find—among many familiar plants—one that had a root "like unto stones" (i.e., which look like testicles). This is one of the "divers kinds of orchis, which are called in Latin *testiculus*, that is, a stone." Other names include *cullions* (from the Latin *culleus*, a bag, i.e., the scrotum, via the French *couillon*). Among the species described by Turner is one now known as the Early-purple orchid (*Orchis mascula*), which "hath many spots in the leaf, and is called adder grasse in Northumberland; the other kinds are in other countries called fox stones or hare stones, and they may, after the Greek, be called dogstones" (i.e., *Cynosorchis*). As to their virtues, Turner's account is extremely familiar:

> They write of this herb that, if the greater root be eaten of men, it maketh men children, and if the root be eaten of women it maketh women children; and moreover this is also told of it, that the women of Thessaly give it with goat's milk to provoke the pleasure of the body, whilst it is tender, but they give the dry one to hinder and stop the pleasure of the body.[2]

These details are, of course, straight from Dioscorides, whose works remained important to Western medicine long after the fall of the Roman Empire. The same details are repeated, more or less exactly, in most of the other early English herbals. For example, thirty years after Turner, John Gerard's *Herb-*

all (1597) acknowledged Dioscorides as his source when he came to explain the uses of orchids. He noted that "our age useth all the kindes of stones to stirre up venerie [i.e., lust]," but he told interested readers that the plant called goat's stones (*Tragorchis*, the species now known as the Lizard orchid, *Himantoglossum hircinum*) worked best. However, those intent on stirring themselves up, should know that "the bulbs or stones are not to be taken indifferently, but the harder and fuller, and that which containeth most quantitie of juice: for that which is wrinkled is less profitable or not fit at al [*sic*] to be used in medicine."[3]

Turner and his contemporaries were not being lazy; for more than a millennium, most Western scholars had been convinced that the philosophers of Ancient Greece and Rome had known everything, but much of that knowledge had been lost during the period that became known (somewhat unfairly) as the Dark Ages. Scholars labored in Europe's universities and monasteries throughout this period, but many saw their main task as being to preserve as much ancient wisdom as possible, including any fresh fragments they could find of lost texts (the excitement generated by a lost manuscript of Aristotle in Umberto Eco's *The Name of the Rose* is a vivid fictional account of a very real passion). As a result, the ancient manuscripts had been copied, amended, and commented upon for fifteen centuries, but the same basic information about orchids remained in each. And what was true of orchids was true not merely of all the other plants, but of much of the rest of Western knowledge. The idea of creating entirely new knowledge—such as discovering completely new lands, people, and things—was almost unimaginable. That began to change radically in the sixteenth century and the new plants described by the Badianus manuscript, including the vanilla orchid, are crucial to understanding how modern Western science—and arguably the whole modern world—came into existence. But before we can understand the revolution in learning that is embodied in that little red book, we first have to understand what had happened to orchids—and to Western science and medicine—since the end of the classical world.

Utopian Botany

The various comments on orchids found in the early modern herbals are entirely typical of botany at this period; translations of and commentaries upon Dioscorides continued well into the Renaissance and the era of the printed book. Nevertheless, despite the reverence for ancient learning—and the degree of stagnation that often resulted—things gradually started to shift. Paradoxically, an event that Europeans regarded as one of the greatest catastrophes in their history, the fall of Constantinople (ancient Roman Byzantium, the city now called Istanbul) in 1453 to the armies of Sultan Mehmed II, may have precipitated a growth in Western learning. Perhaps because they were rescued by fleeing scholars, Greek manuscripts started arriving in Western Europe from the former Eastern Roman empire. And many other books came from the Muslim world, where they had long been preserved but were inaccessible to scholars who could not read Arabic. Taking advantage of the newly arriving manuscripts, Pope Nicholas V, a major patron of learning, commissioned Latin translations of Theophrastus's *Enquiry into Plants* and *The Causes of Plants*. They were completed by 1454: the first time anyone in Western Europe had read Theophrastus directly in over a thousand years.

At around the same time that Constantinople fell, printing with movable metal type was perfected in Mainz, Germany, and an explosive production of books began; it has been estimated that in their first fifty years (ca.1450–1500), European printers produced more books than the whole of humanity had produced in its entire history up until this point.[4] The ancient botanical classics must have been in considerable demand, since they were among the titles produced by the famous Venetian printer Aldus Manutius (ca.1450–1515). He produced a Latin edition of Dioscorides (1478), a Greek edition of Theophrastus's works (1497), and then a Greek edition of Dioscorides (1499).[5] These Aldine editions, as they were known, were small, pocket-sized books. Printing made books cheaper, but the Aldine editions made them really cheap, beginning the process that would eventually make books available to everyone. (To help keep his books small, Aldus commissioned a beautiful new typeface, modeled on the handwriting of contemporary scholars, that took up less space; since it originated in Italy, the type became known as *italic*.) Such was the fame of the Aldine books, that when Sir Thomas More described his imaginary island of Utopia, he depicted his fictitious traveler Hythlodaeus using the Aldine edition of Dioscorides (which listed plants alphabetically) as a dictionary from which he taught the Utopians to read Greek.[6] More's fantasy was, in part, an imaginative reflection on the significance of America's discovery, and giving the utopians classical languages and learning mirrored a real spread of learning that was repeated across Europe as more and more of the ancient texts appeared in new, affordable editions, creating a utopia for book-hungry scholars who had relied on hand-copied manuscripts ever since writing had been invented.

However, the spread of printing was not all good news. The increasing availability of texts allowed scholars to compare different versions and they started to notice mistakes, especially in Pliny whose command of Greek appears to have been rather less than perfect. An Italian doctor, Niccolò Leoniceno, published *De Plinii et Plurium Aliorum Medicorum in Medicina Erroribus* (On the Errors in Medicine of Pliny and Many Other Medical Practitioners, 1492), which identified the many mistakes Pliny had made, usually mistranslations from Greek. Other scholars either joined in the attack or defended Pliny, arguing that his mistakes were really those of later copyists. These debates gradually created a widespread recognition that the ancient texts were far from perfect. Leoniceno was among those who argued that the centuries-old practice of simply comparing texts could never resolve these arguments; what was needed were fresh observations, new first-hand information about the natural world that could be used to produce what he called a natural history written "not from words, but from things." The ancient sources were not, of course, abandoned. On the contrary, it was their careful re-reading that led to a fresh emphasis on first-hand observation in the early sixteenth century.[7]

This desire to supplement book-learning with new facts was given a dramatic impetus, in the same year that Leoniceno published, when Christopher Columbus reported the discovery of new lands full of extraordinary novelties. "It grieves me extremely," Columbus wrote of the new animals he had found, "that I cannot identify them, for I am quite certain that they are all valuable and I am bringing samples of them and of the plants also."[8] Columbus, of course, initially believed that he had reached the shores of Asia, but his fellow Italian, Amerigo Vespucci, showed that the new continent (to be named in his honor) was indeed a whole New World, whose existence had apparently never even been suspected by the great thinkers of antiquity. The discoveries of Columbus, Vespucci, and those who followed them led to the creation of a new genre of books that stressed novelty rather than tradition, reveling in the opportunity to describe and name countries, peoples, animals — and plants such as vanilla — that no European had seen before.

Over the next century, there were numerous attempts to catalogue and name the wonders of the New World. In 1570, Francisco Hernández, physician

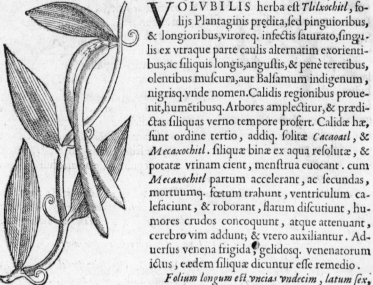

38   RERVM MEDICARVM NO. HISP.

De T L I L X O C H I T L seu flore nigro Araco Aromatico. *Cap. X V.*

VOLVBILIS herba eft *Tlilxochitl*, fo-
lijs Plantaginis prædita,sed pinguioribus,
& longioribus,viroreq. infectis saturato,singu-
lis ex vtraque parte caulis alternatim exorienti-
bus;ac siliquis longis,anguftis,& penè teretibus,
olentibus muscura,aut Balsamum indigenum,
nigrisq.vnde nomen.Calidis regionibus proue-
nit,humetibusq.Arbores amplectitur,& prædi-
ctas siliquas verno tempore profert. Calidæ hæ,
sunt ordine tertio, addiq. solitæ *Cacaoatl*, &
Mecaxochitl. siliquæ binæ ex aqua resolutæ, &
potatæ vrinam cient, menstrua euocant . cum
Mecaxochitl partum accelerant, ac secundas,
mortuumq. fœtum trahunt, ventriculum ca-
lefaciunt, & roborant, flatum difcutiunt, hu-
mores crudos concoquunt, atque attenuant,
cerebro vim addunt; & vtero auxiliantur . Ad-
uersus venena frigida, gelidosq. venenatorum
ictus, eædem siliquæ dicuntur effe remedio .

*Folium longum est vncias vndecim , latum sex,
siliqua vero longa vncias sex,crassa digitum vnum .*

Figure 7. The Black Flower (*flore nigro*) or Tlilxóchitl (Vanilla).
(From Francisco Hernández, *Rerum medicarum Novae Hispaniae thesaurus*, 1651)
Credit: Getty Research Institute, Los Angeles (93-B9305).

of the Spanish King Philip II, set sail for the colony of New Spain (as Mexico
was then known), with the grand title *Protomédico de Indias* (chief physician of
the Indies) on a seven-year mission to seek out new plants and new animals, to
boldly describe what no European had described before. Hernández learned
the indigenous people's language, Náhuatl, and tried to learn what they knew.
He even translated what he discovered back into Náhuatl, so that New Spain's
inhabitants could read and profit from it. By the time he returned to old Spain
in 1577, he had several shiploads of the New World's plants and thirty-eight volumes of notes and drawings. Sadly, Philip II seems to have lost interest in natural
history by this time, or perhaps Hernández's friendships with various freethink-
ing humanists earned him the unwelcome attention of the Inquisition. Whatever
the reason, most of Hernández's work was still unpublished when many of his
manuscripts were destroyed by a fire in 1671.[9]

That little red book from the Vatican library, the Badianus manuscript, had been written by a Náhua healer called Martín de la Cruz (and hence is also known as the *Codex de la Cruz Badiano*) and it did not fare much better than Hernández's manuscripts had.[10] The book was produced less than thirty years after the fall of Tenochtitlan, by which time the Spanish had created an alphabetic version of Náhuatl and taught a whole generation of Náhuas to read and write it. Among them was de la Cruz, who used the new script to record what he knew about his country's plants and their traditional uses. Juan [sometimes Joannes] Badiano, another indigenous man, was employed to translate the work into Latin (he was a professor at the local Colegio de Santa Cruz) and gave it its Latin title (*Libellus de Medicinalibus Indorum Herbis*, The book of Indian medicinal herbs, 1552). The Spanish had imposed Latin as the language of medicine and learning, but in any case, the book was intended as a gift for the Spanish king, who would not have been able to read the language of his newly conquered subjects. The finished manuscript included almost 200 beautiful colored illustrations of Mexican plants and was kept at the Spanish royal library at the Escorial (the great palace outside Madrid). In the seventeenth century it became part of the library of Cardinal Francesco Barberini (and hence acquired yet another name, the *Codex Barberini*) and it was from there that it moved to the Vatican library. Only when it was rediscovered in the 1920s was an English translation produced, and the book began to be widely studied. It was initially hailed as the only "Aztec" herbal known not to be "contaminated" by European medical and herbal lore, but later scholars dismissed it as derivative, a slavish imitation of European herbals of the time. The reality lay somewhere between these claims: the indigenous team clearly knew European herbals (at one point, Pliny is cited as an authority), but recent scholarship suggests that the book contains a complex mixture of indigenous and European knowledge. For example, many of the illustrations incorporate glyphs (the traditional symbols used to write Náhuatl) and there are other traces of pre-conquest beliefs and practices scattered through the book.[11]

It might seem surprising that the Catholic priests who ran the Colegio de Santa Cruz should have been training people they saw as heathens to record beliefs that the Spanish must have regarded as superstitions, but in that same year that Columbus arrived in America, a new Spanish pope, Alexander VI, set in motion a short-lived reversal of the Church's unrelenting hostility towards magic. The Renaissance generally saw a renewed interest in what was called natural magic and the recovery of ancient, occult texts. For some Europeans, the New World and its fabulous new species generated fresh interest in the idea of controlling nature through natural magic and so, far from ignoring or destroying Native American magical knowledge, many Europeans sought it eagerly. Hence in the 1540s, the Franciscan missionary Bernardino de Sahagún told his students—including Martín de la Cruz—at the Colegio de Santa Cruz at Tlatelco to start recording everything they could learn about indigenous plant lore, medicines, and remedies.[12]

Although some of the key manuscripts describing the New World's plants were lost or forgotten for centuries, many others did reach a Renaissance readership. Carolus Clusius (or Charles L'Écluse) translated works published in languages such as Portuguese or Spanish, which few other Europeans could read, into Latin (still the language of learning across the continent at that time). Among them was Nicolás Monardes' book on New World medicines, which became in Latin the *Simplicium Medicamentorum ex Novo Orbe Delatorum* (Medicinal simples [herbs] from the New World described), which was printed in Antwerp by another of the great Renaissance printers, Christopher Plantin, in 1579. The book was then translated into many vernacular languages, including English (in which it became *Joyful Newes out the New Founde Worlde*).[13]

Clusius had been a member of the courts of the Holy Roman Emperors, Maximilian II and Rudolf II (in Vienna and then briefly in Prague), both of whom were fascinated by the New World's many wonders. With their help, Clusius amassed considerable collections that he took back to the Netherlands when he became a professor at the University of Leiden. As part of the Europewide trend towards supplementing book learning with direct observations, many universities were founding physic or medicine gardens (the forerunners of botanic gardens) that were used to teach their students the medical properties of plants. So, universities started hiring distinguished naturalists and the professors of Leiden (founded 1575) persuaded Clusius to help develop its botanical garden. He grew whatever exotic plants he could obtain, including species from the Dutch East Indies (modern-day Indonesia) that he obtained with the help of the Dutch East India Company. However, their ships' captains were not as helpful as the Holy Roman Emperors had apparently been, so he began to develop extensive correspondence networks with scholars, merchants, and travelers to acquire both specimens and information.[14]

The culmination of Clusius' work was the *Exoticorum Libri Decem* (Ten

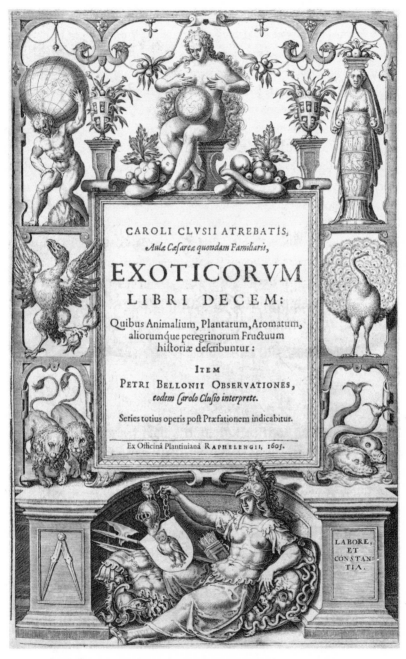

CAROLI CLVSII ATREBATIS,
Aulæ Cæfareæ quondam Familiaris,

EXOTICORVM
LIBRI DECEM:

Quibus Animalium, Plantarum, Aromatum,
aliorúmque peregrinorum Fructuum
hiſtoriæ deſcribuntur :

ITEM
PETRI BELLONII OBSERVATIONES,
eodem Carolo Cluſio interprete.

Series totius operis poſt Præfationem indicabitur.

Ex Officinâ Plantinianâ RAPHELENGII, 1605.

LABORE,
ET
CONSTAN-
TIA.

Figure 8. Clusius' *Ten Books of Exotic Things* described many new plants and
animals from the Americas. (From *Exoticorum libri decem,* 1605)

Books of Exotic Things, 1605), a massive compilation of everything he had been able to learn about newly discovered plants and animals, many of which came from the Americas. Among the many "exotic things" described were some seed pods that Clusius called *Lobus oblongus aromaticus*, a plant he had acquired from Hugh Morgan, pharmacist to Queen Elizabeth I. These unappealing looking dried black things were the source of a flavoring that the Aztecs used, vanilla (which is why the plant was called the black flower — "black seed pod" would have been more accurate).[15] New World orchids had finally made it into print (even though the species that produced the pods had not yet been recognized as an orchid).

Clusius encapsulated much that was new and distinctive about natural history during the Renaissance. He was widely traveled (although only within Europe) and well connected (princely courts were a crucial source of patronage), he collected extensively (the Renaissance collections known as cabinets of curiosities were the forerunner of today's natural history museums), and he supplemented and amended the knowledge of classical writers with new information drawn from reliable accounts of first-hand observations, and with observations made in his botanic garden. He was among the first to realize that America contained an entirely new flora unimagined by Theophrastus and unknown to Dioscorides; this New World would require new names.[16]

The Signature of All Things

The arrival of printing allowed books like Clusius' *Ten Books of Exotic Things* to circulate more freely, but did not automatically lead to new knowledge. Indeed, one of the unanticipated consequences of the printing revolution was that errors could be disseminated just as rapidly as new facts. Perhaps the most startling example is an edition of the Bible produced by the royal printers in London in 1631. Unfortunately, either their proofreading or their morals were imperfect and the book was circulated with the word "not" missing from the seventh commandment, leaving it to read "thou shalt commit adultery." The edition (which became known as the Wicked Bible) was immediately seized and most copies were burnt, so it is now a very valuable collector's item.[17]

Ethically suspect typos were comparatively rare; much more common was the uncritical repetition of ancient information, such as the endlessly reprinted details of the orchid's aphrodisiac properties that we have seen. However, one significant innovation did occur in sixteenth-century German herbals, when naturalistic woodcut illustrations were added to the written plant descriptions to make plant identification easier. The text was often still taken from Dioscorides, but the pictures were new and became an important feature of the works of three sixteenth-century Germans: Otto Brunfels; Leonhart Fuchs; and Jerome Bock (who Latinized his name to Hieronymus Tragus). Another change was introduced in Bock's new herbal, the *Kreütter Buch* (1539): it was written in plain everyday German — at a time when most learned men still wrote in Latin — and his plant descriptions were simple and elegant, all of which made the book a great success. New editions — to which illustrations were added — were produced in 1546 and 1551 and the following year the book was translated into Latin to increase its circulation beyond the German-speaking lands.

As in most herbals, Bock classified plants according to their medicinal or similar properties, such as strong-smelling plants or those with edible roots, but he also launched a quiet revolution in the study of plants; he made a conscious effort to group similar-looking plants together (instead of grouping plants by human need, such as by the part of the body they were used to treat). Bock described his new groups as having been joined together by Nature, since they often had similar properties, and he thus began a quest to identify the natural patterns that exist independently of human interests. The key to this, he believed, was careful observation so he looked very closely at the plants he studied, often noting features others had overlooked. And in his search for the hidden order of nature, which he assumed was God's original plan for the creation, he looked for evidence of signs or meanings hidden in the plants.[18]

When it came to orchids, Bock included only plants with the characteristic, testicle-like tubers; their flowers were still not used for classification, but he did examine the orchids' blooms and puzzled over how they reproduced, since they did not appear to have seeds. He knew that as orchids withered away each autumn, they produced a very fine dust that fell from where the flowers were, but was convinced that this powder perished along with the plant's stem and leaves.[19] (It was only centuries later that botanists realized that this "dust" was in fact the orchid's minute and numerous seeds.) For Bock, the great mystery was, where did new orchids come from? This was a mystery that would puzzle students of orchids for many centuries to come, but Bock's solution to it added a fresh layer of intriguing mythology to the story of the orchid, an attempt to explain the puzzling resemblances to insects that many orchid flowers exhibit.

As mentioned in the previous chapter, the testicular shape of orchid tubers suggested to the early Greek rhizotomai that the plants must have sexual properties. The idea that the appearance of plants must indicate their uses would become known as the doctrine of signatures and over the centuries that it persisted, it shaped European understandings of orchids (and of many other plants). The fully developed form of the doctrine can be traced to the extraordinary figure of Philippus Aureolus Theophrastus Bombastus von Hohenheim, who gave himself the time-saving Latin name of Paracelsus, which embodied the boast that his medical knowledge transcended that of the great Roman physician Aulus Cornelius Celsus ("para," in this context, meaning "going beyond," as in parapsychology). Paracelsus was briefly professor of medicine at Basel, but he was a difficult, argumentative, and uncompromising man. He decided to lecture in the vernacular Swiss-German (not Latin as all other professors did), and to ignore the classical authorities, using his own experience as the basis of his teaching. He even burnt the works of some of the great classical medical authorities. The students protested even more vigorously than the faculty and he had to leave Basel in a hurry, leaving an unsettled lawsuit and all his manuscripts behind.[20]

Paracelsus, like many of his contemporaries, was a strong believer in alchemy (his father had also been both an alchemist and a doctor). Alchemy was a complex set of ideas about the hidden powers and possibilities for transformation that were common to minerals and living things and Paracelsus, in his search for occult secrets that could cure disease, revived the ancient doctrine of signatures, seeing it as a guide to the unseen powers of plants. A seventeenth-century medical book offers what claims to be a translation from Paracelsus as explanation of the idea:

> I have oft-times declared, how by the outward shapes and qualities of things we may know their inward Vertues, which God hath put in them for the good of man. So in St. Johns wort, we may take notice of the form of the leaves and flowers, the porosity of the leaves [which] signifie to us, that this herb helps both inward and outward holes or cuts in the skin . . .[21]

Unsurprisingly, Paracelsus recommended orchids (which he called *Satyrion*) to arouse desire.[22]

Paracelsian ideas were adopted and extended by Giambattista della Porta (also known as John Baptista Porta), who was born in Naples probably shortly before Paracelsus' death. Porta shared the renewed interest in magic with many other Renaissance thinkers, and wrote:

> I think that Magick is nothing else but the survey of the whole course of nature. For whilst we consider the Heavens, the Stars, the Elements, how they are moved, and how they are changed, by this means we find out the hidden secrecies of living creatures, of plants, of metals, and of their generation and corruption.[23]

Porta analyzed the "hidden secrecies" of plants in a book called *Phytognomonica* (1588), in which he asserted that yellow plants could cure jaundice, while plants with insect-shaped flowers could cure insect bites. In his *Magiae Naturalis* (Natural magic, 1558) he prescribed orchids as an aphrodisiac, arguing that they not only increased potency and prolonged intercourse, but excited women too (lettuce, on the other hand, was to be avoided). The political philosopher, Niccolò Machiavelli shared this widespread belief in the power of orchids; in one of his plays the main character, an elderly libertine, plans a dinner of onions, fava beans, rare pigeon, spices, and orchid roots as an ideal preparation for the planned seduction of a much younger woman.[24]

A typical late-sixteenth-century Paracelsian was the German alchemist and medical professor Oswald Croll (who Latinized his name to Oswaldus Crollius), a university-educated physician who traveled widely but eventually settled in Prague, where he moved in courtly circles and was regularly consulted by the Holy Roman Emperor Rudolf II. Croll's only book was *Basilica Chymical* (Chemical basilica, 1609), but it was very influential, running through eighteen editions in less than fifty years, including French, English, and German translations. The book included a section *De Signatura Rerum* (Treatise of signatures), because Croll explained "God hath written with his own sacred finger" upon all Earthly things, including "Vegetables, Mineralls and Animalls." He criticized herbalists who used only external appearances to assess signatures, arguing instead that a true doctor must be an alchemist too and understand the internal virtues of a plant through true natural magic, the alchemist's art. However, he does not seem to have practiced what he preached when it came to plants, his text being largely compiled from earlier ones.[25] His treatise is organized according to the parts of the body so, for example, under "Of the head" he notes that "Poppy with a crown represents the Head and Brain; therefore a decoction thereof, in many affects [sic] of the Head, is profitably exhibited."

When it came to the signatures "of the privities" [i.e., the genitals], Croll asserted that Pythagoras would not eat beans because they were aphrodisiacs, and "have also the intire Anatomy of the Privities, and glande of the Yard" (a common seventeenth-century euphemism for the penis — and perhaps a monument to male optimism). And among the other supposed aphrodisiacs, we naturally find — under the heading "Of the stones or genitalls":

all the species of Orchis, from their similitude of the Testicles, are exciters of the Venerial [lustful] Faculty, where it is defective: one is dissolved in the liquor of another, the Superiour is greater, and fuller, and is powerful in provoking Copulation; the inferior is softer, and withered, inhibiting the Procreative Faculty. Nature industrious in the Generation of Mankind, by this representation signifies, that these are powerful in Venerial Vertues, Conception and Off-spring.

The idea that nature was "industrious in the Generation of Mankind" typifies a widespread belief that signatures were evidence of God's benevolence not only in providing cures for humankind's ills, but also in guiding us towards them. Those having difficulty in obeying the command to be fruitful and multiply could turn to the flowers of the *Cynosorchis*, which "invite Men to Pleasure, and Lasciviousness, inciting, provoking, and encreasing Venery." Despite his rhetorical demands for detailed alchemical analyses of plants, Croll notes of *Satyrion erythronium* ("that is, red Satyrion") that it "powerfully excites Lust" even if "being only held in the Hand, and more if drunk in Wine, as Dioscorides, and after him Lobelius, testifie."[26] The reference to Dioscorides once again reminds us how long the Roman doctor's reputation survived.

The doctrine of signatures was widely believed in the sixteenth century and the Christianizing of what might have once been dismissed as pagan superstitions helped make it acceptable. The astrological botanist Robert Turner boldly asserted that "God hath imprinted upon the Plants, Herbs and Flowers, as it were in Hieroglyphicks, the very signature of their vertues." And among the later proponents of signatures was William Cole, a fellow of New College Oxford,

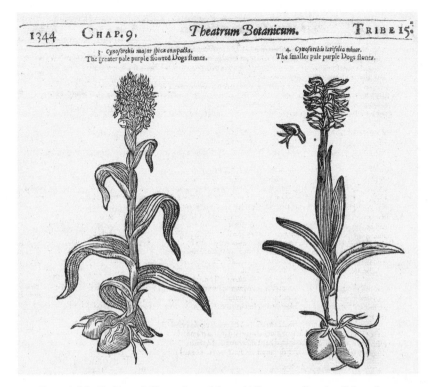

1344   CHAP. 9,   *Theatrum Botanicum.*   TRIBE 15:

3. *Cynosorchis major spica compacta.*
The greater pale purple flowred Dogs stones.

4. *Cynosorchis latifolia minor.*
The smaller pale purple Dogs stones.

Figure 9. John Parkinson's illustrations of the orchid's supposedly aphrodisiac tubers.
(From *Theatrum Botanicum*, 1640).

who published a book called *Adam in Eden* (1657) in which he explained how the doctrine could be applied to medicine. He noted, for example, that a walnut's wrinkled kernel strongly resembled a brain and so it must be "exceeding good for wounds of the head." However, Cole—like others before him—had to account for the puzzle that some plants that were known to possess medicinal properties seemed to have been left "unsigned" by their creator. He suggested that God intended us to note the clue embodied in those plants that bore signatures and exert ourselves, working diligently to explore creation in search of other remedies. Besides, if all were signed "the rarity of it, which is the delight, would be taken away by too much harping on one string."[27]

However, by the time Cole published, the doctrine of signatures was starting to fade. Thirty years before Cole, John Parkinson, apothecary to both kings

James I and Charles I, had expressed some doubts about the aphrodisiac properties of orchids in his book *Paradisi in Sole Paradisus Terrestris, or, a Garden of
All Sorts of Pleasant Flowers* (1629):

> for force of Venereous quality I cannot say, either from my selfe, not having
> eaten many, or from any other, on whom I have bestowed them . . . It would
> seeme, that Dioscorides doth attribute a great Venereous faculty to the seede,
> whereof I know not any hath made any especiall experiment with us as yet.[28]

The phrase "especiall experiment" and his admission that he had not "eaten
many" himself are both evidence of a new more critical attitude that characterized what is usually called the "Scientific Revolution" (even though its precise nature and scope is much disputed by historians).[29] The term is most associated with mid-seventeenth-century astronomers like Galileo Galilei, Johannes
Kepler, and Isaac Newton, but Parkinson's comments show that what was being
called the experimental philosophy, knowledge built on first-hand evidence and
proper experiments, was spreading into the natural history sciences too. However, the term "revolution" shouldn't lead us to think that the changes were
rapid, nor that old knowledge was simply discarded. A decade after his skeptical
comments on orchids, Parkinson wrote a much longer book, *Theatrum Botanicum* (Theatre of Plants, 1640), in which his comments sound rather more old-
fashioned. He noted that "Pliny also writeth the same words out of Dioscorides,
yet it is generally held, but almost all now adayes, that the firme roote onely is
effectuall for that purpose, and the loose or soft spongy roote to be either of no
force or to hinder that effect." As a founder of the Society of Apothecaries, which
made an early attempt to regulate the trade in drugs, Parkinson was contemptuous of some of his contemporaries who pretended to an expertise they lacked;
even though the different powers ascribed to the orchid's two tubers had supposedly been known since ancient time, "most of our Apothecaries doe promiscuously take, not onely both of those rootes to use, but of all sorts of Orchides
in generall."[30]

A decade later, one of the most widely read early English herbals, Nicholas
Culpeper's *Complete Herbal* (1653), simply copied quite shamelessly from
well-known sources. Among these was the well-known *Herball* by the London
apothecary John Gerard, and Culpeper simply repeated Gerard's caution that
the roots of orchids "are to be used with some discretion"; but as the doctrine

of signatures had largely fallen from favor by this time, Culpeper offered an astrological explanation for the plant's power, explaining that the roots "are hot and moist in operation, under the dominion of Venus, and provoke lust exceedingly." By the time Culpeper wrote, the doctrine of signatures had been losing ground for over half a century. The Flemish botanist Julius Rembert Dodoens (1517–85) asserted that "the doctrine of the Signatures of Plants has received the authority of no ancient writer who is held in any esteem: moreover it is so changeable and uncertain that, as far as science or learning is concerned, it seems absolutely unworthy of acceptance."[31] However, even as botanists began to abandon the idea that God had hidden medicinal messages in the appearances of plants, the flowers of the orchid were being interpreted as evidence for an even stranger idea.

Bock may have been the first to write about the curious resemblance between orchid flowers and insects or birds. In his detailed discussion of the strange-looking Bird's Nest orchid (*Neottia nidus-avis*), he remarks that it grows in the hedges and bushes where small birds mate and nest.[32] The orchids, he concluded, grow where the semen of such birds has fallen on the ground. Although Dodoens was dismissive of signatures, he took up Bock's ideas and found numerous examples of insect (and other forms of) mimicry among orchids. Although he was unable to make any progress in understanding the cause of these bizarre resemblances, his close and careful observations would stimulate later botanists to look ever more closely at flowers. (Dodoens' chapter on orchids also holds an important place in botanical history because it was the first with illustrations of enlarged details of flowers.[33] The use of such close-ups rapidly became a standard part of botanical illustrations, allowing the anatomy of plants to be more fully understood.)

The idea that orchid flowers sprang from the semen of the insects they resembled survived for over a century after Bock first proposed it. It was taken up by the Jesuit naturalist Athanasius Kircher, who possessed a cabinet of curiosities in Rome, full of remarkable wonders that included machines powered by water or magnets, one of which animated a statue to create a moving picture of Christ's resurrection.[34] Among the more than 30 books Kircher published was *Mundus Subterraneus* (The subterranean [hidden] world, 1665), which in some ways continued the earlier alchemists' quest for Nature's hidden powers and secret forces. Among the many other marvels discussed were those orchids that resembled insects. Kircher argued that just as bees and wasps were spontaneously generated from bull's and horse's carcasses (respectively), so the bee- and wasp-like orchids must spring from the ground wherever the respective animal's semen fell.

It is not clear how seriously Kircher intended these claims to be taken; his book described machines and experiments that did not work and could not have worked, so perhaps it was a kind of philosophical jest, intended to provoke new ideas rather than describe well-founded ones (his suggestion about the orchid flowers does not seem to have been taken too seriously). Yet a decade later, and a century after Dodoens, a Flemish merchant called Jakob Breyne was still intrigued by the mimicry of orchids:

> If nature ever showed her playfulness in the formation of plants, this is visible in the most striking way among the orchids. The manifold shape of these flowers arouses our highest admiration. They take the form of little birds, of lizards, of insects. They look like a man, like a woman, sometimes like an austere, sinister fighter. Sometimes like a clown who excites our laughter. They represent the image of a lazy tortoise, a melancholy toad, an agile, ever chattering monkey. Nature has formed orchid flowers in such a way that, unless they make us laugh, they surely excite our greatest admiration. The causes of their marvellous variety are (at least in my opinion) hidden by nature under a sacred veil.[35]

That "sacred veil" was not to be lifted until the twentieth century. A great deal of work would be needed before the mystery of orchid mimicry could be solved, starting with the most important—but humble—step in the history of any science; naming. Two thousand years after the name orchis had first appeared, nobody suspected that the European plants with their testicular bulbs were in any way related to New World plants such as vanilla. Orchids, in the modern sense of a family of related plants, still did not exist.

3

The Name of the Orchid

For two thousand years after Theophrastus first recorded their name, nobody was quite sure what to call orchids and by the end of the eighteenth century, the need to finally sort out their names had become urgent. To understand why, we need to look at the history of plant classification, a story that follows Europeans' gradual recognition of just how many orchids there were. Dioscorides had named about 500 plant species of which just two were orchids; fifteen hundred years later, Europeans had been gathering and naming orchids for centuries but had accumulated a grand total of only 13 species.[1] Yet, as we saw in the previous chapter, even this tiny group already had far too many names. In Britain alone the few known species of orchids had dozens of different common names—Adder Grasse, Dog's Stones, Dead Man's fingers, or Goat's stones. Across Europe, each country added many more names for the same plants; one orchid was known in Italy as *Uomo nudo* (Naked man), in Germany as *Italienisches Knabenkraut* (Italian Lad's weed), and in France as *Orchis ondulé* (Wavy orchid). And, as if the picture weren't complicated enough, Europe's learned botanists had confused it further by adding various Latin and Greek names, from the original *Orchis* to *Satyrion*, *Cynosorchis*, *Serapias*, and *Tragorchis*, but there was no unified system for using these names (or checking whether someone else had already used them). However, the real problem was the vagueness of all these names: one species might have many different names (often within a single country), but even more confusingly, many different species might be known by the same name. Whenever Europe's herbalists, botanists, alchemists, explorers, apothecaries, and merchants wrote or talked to each other about plants, endless confusion resulted.

However, the chaos of Europe's plant names was nothing compared with what was to happen in the centuries after Columbus and Vespucci; the Náhuatl name for vanilla, *tlilxóchitl*, embodied an enormous threat to the stability and usefulness of European names not merely for orchids, but for all plants. Europe holds at most one per cent of the world's orchids, yet—as we have seen— naturalists had tied themselves in nomenclatural knots over fewer than a dozen species. As Europeans explored, traded with, and conquered more and more of the world beyond their shores, many new species of orchids would gradually be found, but they were just a tiny fraction of the ever-greater numbers of unfamiliar plants being discovered, for which Europe had neither classical nor vernacular names. At times, Europe's naturalists must have felt as if every ship arriving from the tropics brought a fresh set of headaches with it; each new species of bird, animal, insect, or plant needed a new name—whether it came from the New World, the East Indies, Africa, or the distant Orient. Europe's two-thousand-year-old tradition of natural history was about to collapse as the treasures of the wider world piled up in nameless heaps.

To make sense of this overwhelming influx of novelties, Europe's botanists started creating new ways of classifying and naming them. As we have seen, the earliest botanical classifications were human-centered, based around the uses to which we believed plants could be put. Theophrastus and a few others had begun studying plants for their own sakes, but the purely practical approach embodied in the medieval and early modern herbals was far more common. Oswald Croll's decision to organize plants as Dioscorides had done—according to the parts of the human body they were thought to cure—was typical of a long-standing tradition that placed humanity, quite literally, at the center of the universe. The ancient astronomers had assumed that the Earth was the still point at the center of a closed cosmos, which evolved into the Christian idea that it had been created by God as a habitat for humanity, so naturally everything revolved around us; during the scientific revolution the central Earth was replaced by the Sun, and the Earth was relegated to the status of one planet among others, orbit-

ing in empty spaces that grew ever vaster with each passing century. As European knowledge of the wider world expanded, naturalists were beginning to see
Europe's animals and plants in a similar way, as peripheral to the rich, tropical
lands that held so many new and extraordinary things and were the true centers
of life's diversity.[2]

In the European struggles to catalogue and name all the new plants that
were arriving, we can see the beginnings of new ways of thinking about them.
Clusius, in his *Rariorum Plantarum Historia* (History of rare plants, 1576), followed an arrangement that had its roots in Theophrastus: in the first section of
his book he described trees and shrubs (i.e., plants with woody stems); then he
turned to plants with bulbs (within which he included orchids); then sweet-
smelling flowers, while the fourth group was those with no smell; the fifth was
devoted to poisonous and narcotic plants; while species that have a milky juice
(together with various plants that didn't fit elsewhere) were placed in a final
group. He was clearly trying to create groups of plants that seemed to belong
together, regardless of what they were used for, but human attitudes and opinions were still strongly in evidence. For example, to justify listing often unimpor-
tant plants such as orchids so prominently, Clusius explained that it was because
many of them "attract and delight the eyes of all persons in an extraordinary degree by the elegance and variety of their flowers, and which therefore ought not
to have the lowest place assigned to them among garland-plants."[3] At the same
time, Clusius' friend Matthias de l'Obel (also known as Lobelius, after whom
the genus *Lobelia* is named) took a somewhat similar approach by basing his
classification on the shape of plants' leaves. He started with the simple, narrow
leaves of the familiar grasses and then listed those, like lilies and orchids, that
have slightly larger, broader versions of the same leaf, before moving on to more
complex leaves. As a result (although the terms still hadn't been coined) he
was classifying monocotyledonous plants (monocots) separately from the dico-
tyledonous ones (dicots), just as modern classifications would eventually do.[4]
Meanwhile, their contemporary Dodoens included several images in his books
that are recognizable as particular species of orchid. However, while the quality
of the pictures improved and the language used to describe plants gradually became more precise, the classifications remained much as they had been for centuries, largely focused on human uses.[5]

At the beginning of the seventeenth century, Gaspard (or Caspar) Bauhin,
director of the Basel botanic garden, thought it might be possible to write a

universal history of plants—the *Pinax Theatri Botanici* (The botanical theatre tabulated, 1623). He hoped that by classifying and describing every single plant species, his book would reveal the patterns that existed in nature itself. Leonhart Fuchs had listed 500 species (virtually the same list as in Dioscorides) in his great herbal (1542), but just 80 years later, Bauhin described 6,000 species. Yet despite this huge growth in numbers Bauhin's *Pinax* was, in many ways, still indebted to the ancient, Western tradition of the herbal. Although some historians have seen elements of modern classification in the way Bauhin organized his enormous list, it is clear that classical ideas dominated—and that they were breaking apart under the weight of new species.[6]

However, the twelve-fold expansion in the number of known plant species that Bauhin had to cope with was only the beginning of a deluge of new species that would finally drown the classical botanical tradition. From the mid-seventeenth century, European exploration—and the resultant flood of new plants—expanded at an ever-faster pace, not least because European gardens began sending out explorers specifically to look for new plants: Paris' *Jardin du Roi* alone sent five main expeditions between 1670 and 1704, and the first volume of London's *Philosophical Transactions of the Royal Society* (1666) included detailed instructions "for the use of travellers and navigators" that described how to collect new and exotic specimens.[7] The result would be ever-greater numbers of unfamiliar plants, including new orchids, for Europe's botanists to name and classify.

Making a Family

As the eighteenth century dawned, John Ray—the first man to study botany systematically at the University of Cambridge—completed his massive *Historia Plantarum Generalis* (General history of plants, 1686–1704). Ray noted that in the fifty or so years since Bauhin published details of 6,000 plants in his *Pinax*, the number of known plant species had more than tripled, growing to over 20,000.[8]

The New World was one major source of this growth; Ray knew Francisco Hernández's work, thanks to his friend Hans Sloane (whose collections formed the nucleus of what would become London's Natural History Museum). Sloane not only owned a manuscript copy of Hernández, but also must have been a source of first-hand information; in 1687, he had spent over a year in Jamaica as personal physician to the island's governor and had collected over 800 plant specimens, "most whereof were New." As Sloane explained, he had made the voyage "to see what I can meet withal that is extraordinary in nature in those places." Medicinal plants were of particular interest, and he believed the voyage promised "to be useful to me, as a Physician; many of the Antient [sic] and best Physicians having travell'd to the Places whence their Drugs were brought, to inform themselves concerning them."[9] In seeking first-hand evidence of where exotic plants and medicines came from, Sloane exemplified the scientific revolution's emphasis on experience, but naturally he blended his own observations with quotes from the works of Hernández, Clusius, and several others.[10]

Although the New World's plants were becoming better known, descriptions of many of them had not yet been published; for example, although Hernández had collected approximately 1,300 plant species in Mexico in the 1570s, it was to be another eighty years before a selection from his manuscripts was finally published in 1651 by the *Accademia dei Lincei* (the Academy of the Lynxes, or Lynx-Eyed), arguably the world's first scientific society, which counted Galileo among its members.[11]

The Lynx's edition of Hernández described many New World plants and animals for the first time, but still arranged them according to European ideas. There are traces of Náhuatl ideas as well as similarities to the Badianus manuscript, but Hernández's editors decided to list his plants according to Dioscorides' ideas about the parts of the body they could be used to treat; while this approach made the New World's plants more useful, it also squeezed them into the Old World's categories, making them seem more familiar than they really were.[12] And, in addition to using European ideas and categories to sort and name American species, many of the new plants that were introduced into Europe (including tobacco, maize, and tomatoes) spread so rapidly that botanists were sometimes confused as to whether or not a plant was really new, or—if new—which part of the faraway, unfamiliar world it had come from. As a result, the novelty of the American flora wasn't fully recognized in Europe until the eighteenth century.

John Ray used the Lincean Academy's plant descriptions (as well as those in many similar volumes) for his huge catalogue, which wasn't completed until 1704—so that can usefully be regarded as the date when many of America's plants finally took their place in the European catalogue of nature. The rapid growth in the numbers of known plant species led Ray's contemporary, the French botanist Joseph Pitton de Tournefort, to argue that the time had come to completely reorganize the way plants were classified. He believed that the species was too narrow a category to be useful and proposed that a broader category, the genus (from the Greek word for a race or stock), would have to be treated as the fundamental unit if naturalists were to avoid being overwhelmed.[13]

Tournefort was not the first to group plants into genera (the plural of genus), but he did define them more clearly than any of his predecessors, using common features of the structure of the flower's petals (the corolla) and the fruit to separate species into coherent groups.[14] In his *Élémens de Botanique* (Elements of Botany, 1694), he grouped genera together into even broader groups called orders (roughly equivalent to what today's botanists would call families), one of which was the orchids, defined mainly by their asymmetrical flowers.[15] For nearly two thousand years, orchids had been a scattered handful of species (only the few that had the characteristic testicle-like tubers had been seen as related); as the eighteenth century dawned, they finally became a recognized family.

Tournefort's system of classification was only one among many, so it took time for the idea that orchids were a family to spread, but that didn't stop the orchid family from growing rapidly. Just as Tournefort was naming the group, the final volume of the massive *Hortus Indicus Malabaricus* appeared, one of the world's first great tropical floras. It was the brainchild of Hendrik Adriaan Van Rheede tot Draakenstein (known as Van Rheede), who was the Dutch Governor of the Malabar region of India (now in the state of Kerala). Deeply interested in botany, he used his authority to organize the production of the *Hortus Malabaricus*. (Publication began in 1678 but was not complete until 1693, by which time Van Rheede had died.) Its twelve substantial volumes were the result of over thirty years of hard work by a team of more than 200 local experts, including several Indian priests and physicians who knew about traditional plant-based medicines, together with four soldiers in the Dutch East India Company's army who produced the book's gorgeous illustrations.

Van Rheede clearly loved the Indian flora, and he commented enthusiastically that on one of his travels he had

> observed large, lofty and dense forests. . . . It was often very pleasant to behold on one tree, leaves, flowers and fruits of ten or twelve different kinds displayed.

Figure 10. The Dutchman Van Rheede published one of the first descriptions of tropical orchids growing on trees. (From *Hortus Malabaricus*, 1686–1693)

And yet they did not harm this tree in any way so that the trunks of such trees were very close to each other and very thick, or at all events they lifted their heads in air to an elegant height of as much as eighty feet.[16]

Van Rheede thus became one of the first Europeans to publish a description of those tropical orchids that grow on trees. The plants were sometimes considered parasites; when Sloane described such orchids in the account of his voyage to Jamaica, he presumed they were a form of mistletoe and misnamed them accordingly.[17] However, Van Rheede understood that "they did not harm" the trees they grew on; they simply used the trees for support and so they gradually became known as "air-plants." Today, botanists and gardeners call them epiphytes (literally an "upon-plant," i.e., one that grows on another). The majority of the world's orchids are epiphytic or lithophytic (i.e., growing on stones, rather than in soil); those with the paired tubers that gave them their original name are a minority of species within the orchid family.

A Second Adam

Tropical orchids like those Van Rheede described were arriving in Europe with ever-increasing frequency. At around the time that he published his volumes, new orchids were being identified in works such as *Phytographia* (1691–1694, 1696) by the English botanist Leonard Plukenet, Royal Professor of Botany and gardener to Queen Mary. And when the German naturalist Engelbert Kaempfer returned from many years traveling in Asia, he published *Amoenitatum Exoticarum* (1712), which added more orchids to European catalogues.[18] These new additions further exacerbated the problem of names, because—despite the efforts of Tournefort and others—there were still many different systems and names in circulation. The chaos increased with each new species, but this was about to change thanks to Carl Linné, the rather scruffy son of a provincial Swedish pastor.

Linné (who is better known by the Latinized version of his name, Linnaeus) trained to be a doctor and then worked for several years in the University of Uppsala's botanic garden, which—like most university gardens—was used to teach botany to medical students. Away from large cities, doctors had to make their own medicines and plants supplied most of the ingredients, so Linnaeus

Figure 11. Some of the first Asian orchid species to reach Europe.
(From Engelbert Kaempfer, *Amœnitatum Exoticarum*, 1712)

would take medical students around the garden, teaching the names of plants as well as how to recognize them and which diseases each was used to treat.

To make his task easier, Linnaeus produced a catalogue of the garden's plants. The earliest versions used Tournefort's system of classification, but in the 1731 version, Linnaeus started introducing his own system, which had been inspired by an article he read about the French botanist Sébastian Vaillant, Tournefort's successor at Paris' *Jardin du Roi* (Garden of the King). In 1718, Vaillant had given a lecture at the garden in which he argued that the plant's stamens and pistils (carpels) were their reproductive organs, directly analogous to the penis and other organs of animals; he even joked that flowers must derive a little innocent enjoyment from reproduction, much as we do. The idea that plants were sexual beings was not entirely new, but was still unfamiliar enough to be rather shocking. However, once his arguments were published in good, scholarly Latin, the idea spread rapidly. It became clear that the sexual organs for flowers were more important than their petals (the parts Tournefort had previously emphasized), so Linnaeus decided the reproductive parts of plants would provide a better basis of classification.

Over the next few years, Linnaeus produced a series of books that began with Vaillant's ideas and elaborated and developed them into a new and comprehensive system of botany. Perhaps Linnaeus' most lasting legacy to science was that he pulled together the ideas of many different naturalists and turned them into a single, coherent system. For example, he took the two-part (binomial) Latin names for species that Bauhin had first used and formalized the rules for coining and applying them. It is thanks to Linnaeus that Latin binomials are still used in science; human beings are *Homo sapiens* because that is the name he gave us.[19]

Linnaeus named almost 8,000 species of plants himself (plus about 4,400 species of animals), and he brought stability to many existing names.[20] For example, he fixed the characters that defined the genus *Orchis* in his *Species Plantarum* (1753) and within a few years the orchid that had previously been known as the Naked man, Italian Lad's weed, or the Wavy orchid was finally nailed down as *Orchis italica*. Linnaeus himself named so many species that one of his contemporaries complained about "the unbounded dominion which Linnaeus has assumed" over the natural world. Just as Adam in the Garden of Eden had given names to all the animals, Linnaeus "considered himself as a second Adam" and renamed them all.[21] The nickname, intended as a criticism, became a compliment; Linnaeus was often referred to in his lifetime as the second Adam.

However, when it came to plants, Linnaeus' most important innovation was not new or even consistent names for individual species; it was the system he used to organize those names. Linnaeus built his system on Vaillant's recognition of the ubiquity and importance of plant sexuality. Since every plant had reproductive parts, Linnaeus simply counted those parts to create broad groups. The numbers of male stamens was used to assign the plant to a class, while the number of female pistils or carpels assigned them to a subsidiary order. (Some plants, such as mosses, did not appear to have visible reproductive parts so

Linnaeus assigned them their own group, the cryptogams, those whose "marriages" were hidden.) The new groups were part of a hierarchy, with kingdoms like plants and animals at the top, species at the bottom, and the intermediate groupings of class and order nested one inside the other. This structure meant that anyone who wished to identify a plant needed only to be able to count in order to be able to home in on the correct name. Linnaeus called his system the *methodus propria* (proper method) of classification, but because it was based on counting each plant's reproductive organs, it soon became famous as the sexual system.

The simplicity of the sexual system meant it would rapidly become the first really global method for identifying plants. All over the world, explorers, colonizers, doctors, and others found it invaluable in helping them decide whether an unfamiliar plant was really new and thus potentially valuable. The sexual system also marked another important step away from purely medicinal plant studies towards studying plants for their own sakes.[22] Linnaeus' works were translated into many different languages and these simple, easy-to-use guides helped Linnaean ideas to spread around the world; an order finally began to emerge from the chaos of the previous centuries, but what kind of order?

Artificial to Natural

Linnaeus himself readily acknowledged that the system he had devised produced artificial categories; an order such as *Monandria* (flowers with a single stamen or, more literally, the one-husband group) contained orchids, Canna lilies (which are not true lilies), as well as succulent, salt-tolerant plants like *Salicornia europaea*, the glasswort (the name comes from the fact that they were burned to make soda ash, used in glass-making). Orchids, Cannas, and glassworts have nothing in common beyond the coincidence of having a single stamen. The orders and classes of the sexual system were therefore of no use in predicting the properties of unknown plants (for example, to help colonists find a local plant from which soda ash could be produced).

In his *Systema Naturae* (System of Nature, 1735), Linnaeus wrote that "no natural system of plants has been constructed up till now," but he hoped to be able to "exhibit fragments of it at another occasion." The idea that there must be a true natural system was based on two things: the assumption that God had created all of nature and must have had a plan in mind when he did so; and, the common sense observation that some plants and animals are obviously more closely related to one another than they were to others. Cats are more similar to tigers than they are to tortoises. And roses are similar to apple trees—both have flattish, five-petaled flowers. By contrast, a sunflower is clearly much more like a daisy than it is like a rose. Such broad groupings seem self-evident, based on the overall similarities between organisms, but finer distinctions are much harder to make. "In the meanwhile," Linnaeus maintained, "artificial systems are entirely necessary as long as we lack a natural one."[23]

Linnaeus clearly thought artificial systems (especially his own) were essential tools for identifying and organizing both familiar and novel plants. However, arriving at the natural system was a longer-term empirical project that depended on gathering more specimens.[24] When Linnaeus did begin to work out a truly natural system, he explained his method using a geographical metaphor: "All plants," he argued, "exhibit mutual affinities, as territories on a geographical map."[25] In other words, the common properties that linked daisies and sunflowers were like the common features of two adjacent countries; similar geography helped produce common crops and forms of agriculture, encouraged trade and travel, and sometimes shaped similar languages and even religions. By analogy, closely related plants could be expected to have similar medicinal or other properties. In an age of rapid exploration and expansion of knowledge, understanding the characteristics of new countries was the key to exploiting their riches and Linnaeus saw the plant kingdom in the same terms. One of the main goals of Linnaean botany was to expand his native Sweden's wealth by reducing its dependence on imports, including the many exotic commodities made from plants such as cotton and tobacco. By mapping and classifying the plant world, he hoped to be able to transplant valuable crops to Swedish soil. As he put it, "If Oaks did not grow in Sweden, and some mortal wanted to get Oaks into [the country], and they then grew here as they do today, wouldn't he serve his country more than if—with the sacrifice of many thousands of people—he added a Province to Sweden?" Sweden had no overseas colonies; renaming the plants allowed Linnaeus to give Sweden a botanical empire to make up for the physical one it lacked.[26]

Linnaeus referred explicitly to the economic importance of naming plants accurately, when he wrote:

The generic name has the same value in the market of botany, as the coin has in the commonwealth, which is accepted at a certain price—without necessitating a metallurgic examination—and is received by others on a daily basis, as long as it has become known in the commonwealth.[27]

In other words, if you knew that a plant had its correct Latin name, you knew immediately what you were getting and what it was worth—that was the great benefit of the Linnaean reform.

However, Linnaeus realized that while his sexual system was useful, only "natural orders have their value with respect to the nature of plants."[28] When he finally started producing his promised "fragments" of the natural order, the groups were no longer arbitrary collections of plants, defined by the single property of having the same numbers of stamens. The new natural orders consisted of plants that were linked by many different common properties; his early examples included palm trees (which he called *Palmae*) and orchids (*Orchidae*).[29]

Linnaeus first summarized the orchids he knew in the *Genera Plantarum* (first edition, 1737), where he described eight orchid genera: *Orchis, Satyrium, Serapias, Herminium, Neottia, Ophrys, Epidendrum,* and *Cypripedium*. The last genus (Lady's- or Venus-slipper orchids) is significant because it had two stamens, while all the others had only one; under the sexual system, *Cypripedium* comes within the order Dyandria (two husbands), while all the others are Monandria (one husband), but Linnaeus clearly recognized that the categories established by his own sexual system violated the naturalness of the orchid group. By the time he published the *Species Plantarum* (1753), the number of orchid species had grown from 38 to 62, including increasing numbers of tropical orchids. Ten years later, when the second edition appeared, it included 102 species of orchids and by the end of his life, he had named 113. Since Linnaeus' day, botanists have refined and improved classification. His original collection is now housed in London at the Linnean Society (confusingly, the society's name omits the first "a" from Linnaean). It contains more than 150 species of orchid, including 44 "type" specimens, which are the original unique specimen that botanists consult if they want to know what that species is.[30] In addition to Linnaeus' work, the French naturalists Bernard and Antoine-Laurent de Jussieu refined and improved the system of plant orders (which we now call families) that Tournefort had begun and in the process they gave us the outlines of the modern Orchid family.[31] Thanks to Linnaeus and his successors everyone today knows exactly what an orchid is and how to name it.

Myths of Orchids

Although Linnaeus was a totally serious, scientific man the language in which he wrote about plants seems a very long way from the dispassionate prose we associate with modern plant science:

> The flowers' [petals] themselves contribute nothing to generation, but only do service as bridal beds which the great Creator has so gloriously arranged, adorned with such noble bed curtains, and perfumed with so many soft scents that the bridegroom with his bride might there celebrate their nuptials with so much the greater solemnity. When now the bed is so prepared, it is time for the bridegroom to embrace his beloved bride and offer her his gifts.[32]

Linnaean botany became increasingly popular, but as it spread, some delicate souls worried that the passionate embraces of the flowers were rather unsuitable for impressionable young minds. Botany was often considered a feminine science, especially suitable for mothers to teach their children, since it was regarded as not too difficult and, in contrast with zoology, did not involve the cruelty of killing animals. Unfortunately for those trying to teach the Linnaean system, most flowers have more than one male stamen and more than one female pistil (carpel). As a result, the acceptably monogamous picture that Linnaeus described — "the bridegroom with his bride" — seldom applies; in most flowers, multiple brides and grooms are to be found cavorting in the most shameless botanical orgies. The *methodus propria* was a most *im*proper method in some eyes.

Linnaeus himself was no libertine (he even forbade his daughters to learn French because he thought the language was associated with loose morality), but other writers saw the love lives of plants in very different terms. When Erasmus Darwin (grandfather to Charles) produced a popular translation of Linnaeus' work in English, *The Loves of the Plants* (1789), he not only acknowledged but celebrated the various non-monogamous arrangements in the plant world. Darwin described sex as "the cordial drop in the otherwise vapid cup of life," a source of the purest happiness to be celebrated (and engaged in) whenever pos-

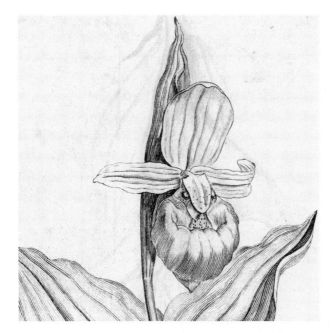

Figure 12. Amorous orchids such as this *Cypripedium* were among the many
Erasmus Darwin described in his poem "The Botanic Garden." (From his
translation of Linnaeus' work into English, *The Loves of the Plants*, Part II, 1791)

sible.[33] And when it came to botany, he took pride in the fact that English was
a more direct, expressive language than Latin, which had allowed him to make
his translations (including their erotic connotations) even more "expressive and
concise" than the originals.[34]

The link between plants and sex, which both Linnaeus and Erasmus Darwin emphasized, was not of course entirely new (although it became both more
widely understood and more significant to science during the eighteenth century). As we have seen, the alleged aphrodisiac properties of plants had been a
source of interest since ancient times, with orchids being one of many types of
plant believed to possess the power to help a man rise to the occasion. When the
Spanish conquistadors met the Aztec emperor Moctezuma (Montezuma), one
of them recorded that he consumed large quantities of a "certain drink made of
cacao, which they said was for success with women."[35] The drink was flavored
with various spices including chili and *tlilxóchitl*, or vanilla, and called *chocolatl*
(from which English derives the word chocolate) in Náhuatl. The use of the

vanilla orchid to flavor *chocolatl* is widely reported in early records of the drink; Hernández lists *tlilxóchitl* among the drink's ingredients and adds that "the property of the drink composed [i.e., by mixing the ingredients] is to excite the sexual appetite."[36] Interestingly, when Hernández lists the properties of *tlilxó-chitl* taken on its own, he makes no mention of it being used as an aphrodisiac; he seems to have believed that it was only when mixed with chocolate that it became so potent. In fact, there's no evidence that the indigenous people thought that either chocolate or vanilla were aphrodisiacs. The Badianus manuscript and other early records of indigenous medicine list vanilla's many medicinal uses: for example, Hernández suggests that vanilla "hastens birth, expels afterbirth and a dead foetus." It could also be used to treat poison and was considered good for stomachache. Vanilla also "expels flatulence" and (rather fortunately, given the previous property) was highly aromatic, but there's no mention of aphrodisiac properties.[37]

Yet by the late eighteenth century many in Europe believed vanilla was an aphrodisiac. The aphrodisiac properties of vanilla were mentioned by the German physician Casimir Medicus in his letters to fellow-doctor Johann Georg Zimmermann *Über einige Erfahrungen aus der Arzenei-Wissenschaft* ("On experience in medical science," Mannheim 1766), in which it was claimed that "no fewer than 342 impotent men by drinking vanilla decoctions, have changed into astonishing lovers of at least as many women"[38] (I love that "at least").There are other eighteenth-century references to vanilla as an aphrodisiac, but little evidence as to when or how this story got attached to the orchid. The story of Moctezuma's drink being an aphrodisiac was probably part of the Spanish propaganda that portrayed him as a proud, lustful, vicious tyrant so that they could justify conquering him (the description of him drinking chocolate comes immediately after one of him eating the most tender parts of young boys and girls, delicacies saved for the king from his people's supposedly daily human sacrifices).

Whatever the indigenous people thought, the Spanish seem to have assumed that the new world would contain new drugs, including aphrodisiacs (tobacco was also considered an aphrodisiac when it first reached Europe). Like most Europeans of the period, the Spanish associated the tropics with exotic sensuality; in their imagination, the foods and spices of newly discovered hot countries were always exciting, often magical.[39] Above all, the tropics were usually associated (in the minds of the men who wrote about them), with fantasies of outlandish sexual encounters with dusky native beauties. So it is not surprising that when some of the first tropical orchids that reached Europe were linked to the exotic flavors of chocolate and vanilla, they were also assumed to be aphrodisiacs.

The aphrodisiac properties of vanilla are not the only example of the way that the eighteenth century adorned the orchids with sexy myths. In 1704, the French writer Louis Liger published a gardening book called *Le Jardinier Fleuriste et Historiographe* (The floristic and historical gardener), in which he recited the ancient myths and legends from which so many popular garden flowers got their names. Liger's book was translated into English in 1706 and explained the origins of the orchid's name:

> *Orchis* was the Man in the World most given to Women. His Father was a Satyr call'd *Patellanus*, and his Mother the Nymph *Acolasia*, who always presided at the Festivals celebrated to the Honour of *Priapus*.

The tale continues:

> 'Twas at the Festival of *Bacchus*, that *Orchis*, like others of the same Character, being in Drink, committed the most disorderly Actions that can be imagined. Being the Son of A Rural God, he thought he might do any thing with Impunity; and his brutish Passion blinded him so far, that he had the Insolence to lay Hands on one of the Priestesses of Bacchus, for which he was punished upon the Spot; for the Priestess did so incense the *Bacchantes* or Assistants at the Festival against him, that they fell upon him and pulled him almost to pieces; and all that his Father could obtain of the Gods, was to have him turn'd into a Flower, which was to perpetuate his Name, as a lasting Stain upon his Memory.[40]

You will find versions of this story in almost every popular book on orchids (and on most websites devoted to them), where it will invariably be described as an Ancient Greek or Roman myth. This seems plausible, since the wine-god Bacchus and his attendants—the beautiful nymphs and lustful satyrs—were usually associated with gardens, fertility, and relaxation and regularly feature in classical friezes and garden frescoes.[41] However, five minutes with any dictionary of classical mythology is enough to throw doubt on the myth's classical pedigree;

the names Orchis, Patellanus, and Acolasia are not to be found in any classical source. There seems to be no trace of them (in any language) before Liger published them in 1704, so he probably invented the whole story. That suspicion is reinforced by the rest of his book, which includes similar myths for many other flowers most of which have no classical precedent either. For example, he claims that antirrhinums, or snapdragons, are named after a character called Anthirrinon, "Son of Priapus and the Nymph Phisia," who was very curious, which led to being killed in a quarrel after which Priapus (the permanently engorged son of Bacchus) had him transformed into the flower that bears his name. Again, there's no classical precedent for any of this; the flower's name is actually derived from the Greek words *anti* (αντι, opposite, or counterfeiting) and *rin* (ριν, nose—as in rhino), because it looks like an animal's mouth; the name made its first English appearance in Turner's *New Herball*, where he recorded its common English name as "calfes snout."

However, while Liger's stories were apparently eighteenth-century inventions, there is something familiar about them. The Orchis story was obviously based on that of Pentheus, the Theban King who spied on a women-only Bacchic festival and was torn limb from limb by the crazed Bacchantes, including his own mother. (This story was told most dramatically by Euripides in his play, *The Bacchae*.) There are also echoes of the story of Hyacinth (*Hyakinthos* in Greek), the beautiful youth transformed by the grieving Apollo into the flower that bears his name, or of Narcissus (cousin of the ill-fated Pentheus), another beautiful young man who becomes a flower.[42] But the most obvious similarity is with the many tales in Ovid's *Metamorphoses* in which people were transformed into plants, such as the story of Apollo and his pursuit of Daphne, who was transformed into a laurel tree to save her from the god's lust. It is not clear which flower Ovid called "hyacinth" (the species we now call *Hyacinthus orientalis* doesn't quite fit his description), and it's been suggested by modern scholars that he may have been referring to a species of wild orchid, yet there's no mention of Orchis, his flowers, or his myth in Ovid.[43]

Why did Liger add spurious myths, usually pastiches of classical ones, to a book that—for the most part—is a straightforward, practical gardening manual, full of details about when and where to plant, and how to nurture various flowers? The sources of information on Liger are so few that it's impossible to be certain, but there may be a clue in the fact that each of his myths is accompanied by a moral. For example, the moral of the tale of Orchis is "Nothing escapes the Vengeance of Heaven . . . it punishes those who are so far from reclaiming, that they still run deeper into vice."[44] Liger's eighteenth-century French readers, being cultured ladies and gentlemen, might have read his book as a reworking, perhaps a parody, of the *Ovide moralisé*, an anonymous fourteenth-century French poem that retold Ovid's tales by turning them into a set of improving fables (although paradoxically, the author occasionally spiced them up with erotic details not found in Ovid).[45]

Whether or not the *Ovide moralisé* was Liger's target, there's no doubt that his well-educated readers would have known that his "myths" were in fact clever, modern pastiches, full of witty—and often rather smutty—jokes. For example, before Orchis meets his fate at the hands of the Bacchantes, Liger recounts that "this young Man's chief study, was to find Opportunities of gratifying his Passion. He loved a Shepherdess Nymph [*Nymphe bocagère*, or wood nymph] called Pornis, by whom he had two Children." It is surely no accident that the nymph's name echoes the French word *pornographie*, which—like its English equivalent—is derived from the Greek, *pornographos* (πορνογραφοσ, writing about prostitutes). Yet it is interesting that Pornis was a wood nymph because most European orchids grow in woodlands, so we have a crumb of useful scientific information embedded in the bawdy story.

The strangeness of Liger's book, particularly its blurring of genres, makes more sense in the context of early eighteenth-century Paris.[46] His brother owned a fashionable café there, on the rue de la Huchette in the Latin Quarter (so called because it was close to the university whose students still studied in Latin). Liger wrote a guidebook for visitors to the city (which he described as "a very delicious abode"), in which he described such cafés as

> the meeting place of *Nouvellistes* [newsmongers] & of some wits who meet here in order to conduct conversations on fine literature. To keep them going, all the substances which are most capable of arousing the ideas under discussion are consumed: coffee [and] chocolate.[47]

Café society centered around literature, particularly poems that ranged from scurrilous and obscene satires to serious debates over philosophical matters, including science. Writers and their audiences would have been familiar with the classical models for such works, particularly the poems of the Roman satirists such as Ovid.[48] In such a society, witty, erotic, pseudo-moralistic fables would have made perfect sense within a book devoted to the enlightened pastime of cultivating one's own garden. Moreover, the audience—as they sipped their vanilla-flavored chocolate—would have considered it gauche to even think of asking an author whether his stories were true, or genuinely classical, or expecting the parodies to be separated out from the properly scientific facts in his text. Cultivated café goers demonstrated their sophistication by simply enjoying the rich ambiguity of the author's concoction, ready to appreciate almost anything, as long as it was entertaining and—above all—clever.

French intellectuals of the time (and since) were somewhat dismissive of the English for being dogmatic and too literal-minded, insisting on unambiguous truths and over-obvious facts, instead of celebrating *politesse* and wit.[49] The English, the Parisian café wits believed, wouldn't recognize a sophisticated French joke if it bit them—and perhaps the fate of Liger's fanciful Myth of Orchis has proved them right. As mentioned, it was quickly translated into English and, shorn of its French context, was simply accepted as a classical tale and repeated verbatim from one orchid book to another, right up to the present day. For example, the British horticultural writer Henry Philips included the tale in his *Flora Historica; Or the Three Seasons of the British Parterre* (1824), from where it was copied repeatedly by other writers. Almost identical versions appear in John Newman's *The Illustrated Botany* (1846), in John Keese's *The Floral Keepsake* (1850), and in Richard Folkard's *Plant Lore, Legends, and Lyrics* (1884). The story is even delicately bowdlerized by the twentieth-century American orchid lover Grace Niles, in her memoir *Bog-Trotting for Orchids* (1904); she metamorphoses Orchis from a drunken, would-be rapist into someone who "failed to observe the rules of politeness while attending a festival of Bacchus, and offended one of the priestesses with his rude behavior."[50] The definitive modern version of the story is in Luigi Berliocchi's *The Orchid in Lore and Legend* (1996), where it is recounted as a genuine, classical myth.[51] Since Berliocchi's version appeared in English in 2000, it has been copied (usually without attribution) repeatedly. Just as the fifteenth-century advent of printing allowed errors to become "facts" that were both trusted and rapidly disseminated, the Internet has encouraged the same process on an even greater and faster scale; at the time of writing, the myth of Orchis—presented as genuinely classical—is to be found on over 20,000 websites.

The endless retelling of the Myth of Orchis over the past three hundred years suggests something more than lazy writers who don't check their sources.

Its popularity surely reflects the way it encapsulates the qualities that Western cultures have come to associate with orchids; Orchis supposedly inherited his father's lust and his mother's delicate beauty. He and his flower embody delicacy and sexiness (their name, after all, refers to the most vulnerable part of the male anatomy, a source of new life that's all-too-easily injured). It is no coincidence that tropical orchids entered the European imagination at the same time as Liger's myth; the epiphytes arrived redolent with exotic heat, the supposed flavoring for the favorite drink of cruel and lustful Aztec emperors, and as perfumed aphrodisiacs, plucked from distant jungles by savage conquistadors. The eighteenth century gave the orchids their rational, scientific name, but did so in the Linnaean language of the marriages of plants. Paradoxically, the supposed age of Enlightenment also saturated orchids in images of sex and (thanks to the fate of Orchis) of death, associations they would never shake.

4

Orchidmania

In 1837, subscribers were excited to receive the first of ten parts of a new book on orchids. They had paid £1.11s (probably the equivalent of over £1,100/$1,700 today[1]) for each part, or £15.15s for the full set. The first installment consisted of five beautifully drawn and hand-colored pictures of Central American orchids, executed by Sarah Anne Drake and other skilled botanical artists. Those wealthy enough to buy the book would have needed purpose-built shelves constructed to hold it because, in order to allow the finished pictures to be life-sized, the book had been printed on elephant-folio paper (27 × 15 inches, 685 × 381mm), the largest sheet a Victorian printing press could handle. When bound together, the ten parts made a vast, unwieldy volume. Alongside the beautiful botanical plates were some witty vignettes, including one by the celebrated caricaturist George Cruikshank. Cruikshank's cartoon (known as "The Librarian's Nightmare") showed a team of workmen with pulleys struggling hopelessly to raise the book upright so that it could be read; it was captioned with a quote from the Greek poet Callimachus: *"mega biblion, mega kakon"* — a big book is a big evil. (Callimachus, who helped catalogue the Great Library of Alexandria, was celebrated for the brevity of his poems.)

The nightmare book was the largest and most expensive book on

Figure 13. The "Librarian's Nightmare" by George Cruikshank.
(From James Bateman, *The Orchidaceæ of Mexico & Guatemala*, 1837–43)

orchids ever produced: James Bateman's *The Orchidaceae of Mexico and Guate-mala.* It took six years to complete. Only 125 copies were produced of this bloated monument to orchidmania, an extraordinary disease that gripped many rich men, particularly in Britain, during the nineteenth century. This obsession with orchids resulted from the paradoxical fact that Europeans had been nam-ing, collecting, and drawing orchids for over 2,000 years; they had, for the most part, failed to grow them. The native European species were not beautiful or glamorous enough and the tropical species initially resisted all attempts to culti-vate them. Pictures were as close to tropical orchids as most wealthy collectors were likely to get.

The first man to persuade a tropical orchid to flower in Britain appears to have been the Quaker merchant Peter Collinson, who dealt in expensive fab-rics, a trade that brought him into contact with everyone from fashionable Lon-don society to coffee-house philosophers and members of the Royal Society, as well as the ordinary sea-captains, who brought him his treasures. He corre-sponded with natural history enthusiasts all over North America from whom he obtained various exotic plants and animals. Collinson obtained an orchid from Providence Island in the Bahamas and, apparently in collaboration with an Englishman called Wager, persuaded it to flower in 1731. A picture of it appeared, under the name *Helleborine americana* (it is known today as *Bletia verecunda*), in the book *Historia Plantarum Rariorum* (1732), by the Cambridge professor of botany John Martyn. (Martyn also founded Britain's first botanical society, which met at the Rainbow Coffee House on Watling Street, and he was responsible for popularizing Tournefort's terminology and ideas about classification in Britain.)

However, Collinson's success was a rare example. His friend Philip Miller, who ran the celebrated Chelsea Physic Garden, seems to have managed to get some species to flower, but he noted in his *Gardener's Dictionary* that the epiphytic species, such as those in the genus *Epidendrum*, "cannot by any art yet known be cultivated in the ground, though could they be brought to thrive, many of them produce very fine flowers of uncommon form."[2] In 1787, however, one of these elusive orchids, *Epidendrum cochleatum*, was persuaded to produce its extraordinary, striped purple, shell-shaped flowers in the Royal Gardens at Kew, and by 1794, Kew had fifteen species of this genus in cultivation.

For most of the eighteenth century, those who lacked Kew's resources or patience were unlikely to have seen a living tropical orchid. Few were brought to Europe and most of those died, partly because the climate was too cold. The need to shelter plants from the cold had been recognized in ancient times and the remnants of an early attempt to solve the problem have been found amongst the ruins of the Roman city of Pompeii; the Romans could not make flat glass sheets, so they used sheets of the mineral mica, which is almost transparent when sliced thinly enough. True greenhouses did not appear until the eighteenth century; the earliest appears to be the one built around 1717 for James Brydges, first duke of Chandos, at his estate at Cannons, Hertfordshire. Like the rest of the estate it was intended to display the duke's enormous wealth ostentatiously but without vulgarity. The greenhouse was designed by the Florentine mathematician, architect, and theorist Alessandro Galilei (from the same family as the noted astronomer Galileo Galilei); at the top was a glazed cupola designed to catch the sun. The principles embodied in the Chandos greenhouse were gradually adopted by opulent homes to create an orangery, a warm building with large windows where tropical fruits were grown for the dinner table, but these were designed to mimic a Mediterranean climate and their dry heat usually killed

Figure 14. Probably the first tropical orchid to flower in Britain, *Helleborine americana*.
(From John Martyn, *Historia plantarum rariorum*, 1728)

tropical orchids. It was only after 1800 that greenhouses began to be used more widely and the tricks for growing tropical plants were gradually discovered.

In the late eighteenth and early nineteenth centuries, a flurry of interconnected changes transformed the world and (as a side effect) Europe's gardens: imperial networks of trade and colonization expanded rapidly, bringing more and more exotic plants and animals to Europe; shipping networks grew and sailing times started to fall; industrialization made iron and glass cheaper (as did the abolition of taxes on glass); there was increased scientific understanding of the climates and conditions under which specific plants would flourish; and, the wealthiest men in Europe became ever-more wealthy and looked for novel ways to spend their money in order to impress their friends, neighbors, and rivals. Together, these changes, which historians still call the Industrial Revolution (despite an uneasy realization that the term oversimplifies some very complex phenomena), created the conditions in which orchidmania would emerge and spread.

Imperial growth opened up inaccessible corners of the world to plant-hunters; local networks of colonists, missionaries, and traders made it easier to recruit indigenous guides and porters, and to obtain information and supplies that allowed expeditions to reach and explore previously un-botanized areas. Improved shipping meant that living exotic flowers could be brought back to Europe to be bought and sold by the flourishing nursery trade (especially after the invention of miniature portable greenhouses known as Wardian cases). Meanwhile, the greenhouses that would hold the new arrivals were being improved. Iron glazing bars slowly replaced the older wooden ones (if the wooden ones were made thin enough to admit sufficient light they could not support large sheets of glass). Iron supports and cheaper, factory-made glass allowed larger greenhouses to be built, which created a demand for ever-larger numbers of showy, tropical blossoms to fill them. The information and specimens brought home by the growing numbers of botanical travelers and explorers allowed leading European men of science to study global patterns of vegetation and link them to information about rainfall, temperature, and soil type; as a result, it gradually became easier to recreate the appropriate conditions in which the newly arrived floral immigrants could flourish. But behind all this was money. Vast fortunes, often built on the scarred and suffering backs of African slaves, brought Europe wealth on an unprecedented scale. After the abolition of slavery in Britain and its colonies, that wealth was invested in new factories and machines, new ships and businesses; wealthy industrialists built extravagant homes and estates, aping — and hoping to surpass — the lifestyles of the traditional landed aristocracy. Beginning in England, thanks to late eighteenth-century Quaker ironmasters who developed cast-iron pipes, moist quasi-tropical heat started percolating into greenhouses; coal, the key to Britain's industrialization, smelted the iron, powered the trains that transported the goods, and produced the steam that heated the greenhouses. The conditions for growing tropical orchids became more common from the 1820s onward. Gradually ever-larger greenhouses were built and elaborate heating systems were designed in an effort to recreate what were imagined to be authentic tropical conditions; stiflingly hot, dripping with moisture, and carefully sealed against the bitter drafts of northern winters.[3]

Orchids were perhaps the most prized of the blooms that were brought from all over the world to fill these new hothouses. Orchid-growing was pioneered by commercial nurseries in and around London: Conrad Loddiges and Sons in Hackney was the first, in 1812, but it was soon followed by others including James Veitch, William Bull, and Hugh Low. In the early years of the trade, most of the plants collected would die on the way home. There are numerous accounts of crates full of orchid plants arriving at London's docks, but when opened there was nothing inside but a reeking pile of blackened, rotting plants.[4]

Finding orchids was only one problem for early nineteenth-century orchid collectors; a much more serious one was that the naval captains and others who usually brought the plants home, seldom knew anything about where or how they grew. Epiphytic orchids were still regarded as a novelty in late Victorian times; one nurseryman's catalogue commented that it was "strange" to realize "that these alluring flowers should, like the telegraph, be a privilege of our own age. The ancients, it is true, noticed a few of the ground-orchids, but an epiphyte they never saw."[5] It took many decades of often disastrous experiments to discover how best to cultivate epiphytic orchids. For example, because they usually grow on trees they were assumed to be parasitic like English mistletoe, just as Hans Sloane had assumed when he first saw them a century earlier. In 1815, the editor of the *Botanical Register* commented that "the cultivation of tropical parasites was long regarded as hopeless" because it was assumed that orchids would grow only on the particular tropical trees they were believed to parasitize.[6] As the epiphytes were commonly known as "air plants," they were sometimes believed not to require any water, with predictably disastrous consequences, while others assumed that their epiphytic growth habit was a temporary one, a survival

The prices are liable to fluctuation according to the market.

Cœlògyne	Each—s. d.		Cymbidium	Each—s. d.		Cypripèdium	Each—s. d.
Massangeana .	. 7 6		× eburneo-Lowianum .	42 0		× Arthurianum pul- ⎫	
ocellata .	. 5 0		elegans (Cyper- ⎫	21 0		chellum . ⎭	
—— maxima .	. 10 6		orchis) . ⎭			× Ashburtoniæ .	. 5 0
pandurata	. 10 6		Findlaysonianum ⎫			× —— expansum	. 21 0
Sanderiana	. 63 0		(pendulum) . ⎭			× —— majus .	. 15 0
speciosa	. 7 6		giganteum .	7 6		× Astræa .	. 42 0
—— major	. 10 6		grandiflorum ⎫	42 0		× Aylingii .	
Swaniana	. 21 0		(Hookerianum) ⎭			barbatum .	. 3 6
tomentosa	. 10 6		× Lowiano-eburneum .	42 0		—— Crossii ⎫	7 6
Veitchii	. 42 0		Lowianum (see fig., p. 6)	7 6		(Warnerianum) ⎭	
See also Pleione.			—— concolor .	. 105 0		—— majus	. 10 6
						—— nigrum .	. 10 6
						bellatulum .	. 5 0
						—— album .	
						Boissierianum.	. 105 0
						Boxalli. See villosum	
						× Bruno .	. 10 6
						× Brysa .	. 42 0
						callosum.	. 5 0
						× calurum .	. 3 6
						× Calypso .	. 7 6
						× caɪdinale (Veitch's ⎫	10 6
						var.) . ⎭	
						caricinum (Pearcei).	10 6
						× Carnusianum .	. 10 6
						× —— Veitch's var. .	42 0
						caudatum .	. 10 6
						—— Lindenii ⎫	
						(Uropedium) ⎭	
						—— Wallisii .	. 42 0
						Chamberlainianum .	5 0
						× Charles Canham	. 10 6
						Charlesworthi.	. 3 6

CATTLEYA ACLANDIÆ.

Figure 15. Orchid prices could fluctuate wildly as more specimens arrived. (From James Veitch & Sons, Catalogues of Plants Including Novelties, 1871–1880)

mechanism in difficult times, and therefore assumed the poor plants would be grateful to be "properly" potted up in peat, soil, or rotting tree bark.[7] Naturally, the epiphytes all died. However, gardeners continued to experiment. In the early 1800s, the elderly Sir Joseph Banks, who had been the botanist on Captain James Cook's voyage to the South Pacific and who went on to become president of London's Royal Society, created ingenious baskets of moss and twigs for the orchids to grow in that proved quite successful.[8]

However, the major problem facing the hapless orchids was the fantasy of the tropical jungle that preoccupied their would-be growers. As noted above, the word originally meant a barren waste-ground, but in the European imagi-

Figure 16. Joseph Banks invented one of the first successful baskets for growing
epiphytes such as *Aerides paniculatum.* (From the *Botanical Register*, 1817)

nation "jungle" came to signify a lush, dangerous forest, choked with rampant
and often poisonous plants, and characterized by oppressive, fetid heat. Harry
Veitch, grandson of the pioneering orchid nurseryman James Veitch, wrote one
of the first histories of orchid-growing in which he described how the Loddiges
nursery turned their greenhouses into "hot steamy places" into which all newly
arrived orchids were consigned: "it was occasionally remarked that it was as dangerous to health and comfort to enter" these greenhouses as it was to visit "the
damp close jungle in which *all* tropical orchids were then supposed to have their
home."[9] In reality, of course, many tropical orchids come from mountainous regions with cool nights and it was only gradually realized that the steam-powered
heat of the British greenhouses was turning them into orchid graveyards.

The losses of orchids were so catastrophic that in the late 1820s, London's
Horticultural Society (founded in 1804—it did not acquire its modern name,
the Royal Horticultural Society, until 1861) decided to make a systematic study

of their growth, which was undertaken by their assistant secretary, John Lindley. At a meeting of the Society in May 1830 he revealed what he had discovered about the conditions under which orchids grew in their native countries. His information was limited and sometimes inaccurate, leading to a continued emphasis on excessive heat without adequate ventilation, but he did at least recognize the need for proper drainage, so orchids would now be roasted rather than boiled. More importantly, he established the basic principle that would eventually lead to successful orchid growing: recognizing the diversity of indigenous growing conditions and recreating them as accurately as possible.[10] For example, orchids that grew in tropical lowlands were planted in a stove (a partially buried greenhouse) that simulated the hot and humid conditions of the tropics; Lindley wrote in *Edwards' Botanical Register* (1835) that epiphytes usually grew in the "damp sultry woods of tropical countries; and accordingly we endeavour in our artificial cultivation, to form an atmosphere for them as nearly as possible that which they would naturally breathe in such stations." However, he was now realizing that while these conditions suited some orchids, "there are others which grow most unwillingly, or scarcely survive under such circumstances."[11]

Bateman's mighty book on the orchids included instructions similar to Lindley's; he noted the need to mimic seasonal conditions by giving the orchids a "season of rest," lowering the temperature in the glasshouse to mimic a tropical cool season. He noted that one "peculiarity" of the Mexican and Guatemalan orchids described in his enormous book was their "being more abundant in the higher latitudes and purer air, than in the hot and pestiferous jungles of the coast." They were, therefore, able to withstand a degree of cold, making them perfect for would-be growers who had previously been put off orchids by the cost of heating greenhouses. In addition to resting their orchids, Bateman advised his readers to give their plants plenty of light, good ventilation, but not too much water. He discussed the rival systems used by some of the country's great orchid collectors, observing that Loddiges nursery maintained theirs at much higher levels of heat and humidity than those favored by Joseph Paxton, head gardener to the Duke of Devonshire at Chatsworth; the London system produced magnificent specimens but also "tends to produce exhaustion" and there were fewer plants in bloom at any given time.[12]

The Blooming Aristocracy

Bateman seems to have been the first to name the madness to which his book appealed, and he commented that what he called "Orchido-Mania . . . now pervades all (especially the upper) classes, to such a marvellous extent." This remark appeared in a book that was beyond the means of most potential orchid enthusiasts, but Bateman observed that "the nobility, the clergy, those engaged in the learned professions or in the pursuits of commerce, seem alike unable to resist the influence of the prevailing passion." While Bateman's list of orchid lovers included those who were newly wealthy, he implied that orchids were natural aristocrats and noted that "the happiness of the community at large" would best be promoted if each class of people tended an appropriate class of plants.[13] This snobbish notion — that the aristocrats of the plant world could only be grown by their human equivalents — would not last long.

John Charles Lyons' *Remarks on the Management of Orchidaceous Plants* (1843) hardly appears like a revolutionary document, but it was the first-ever European manual on orchid growing and helped break the upper classes' rather exclusive hold on the orchid. Lyons was a member of the landed gentry who farmed at Ladiston (or Ledestown), his family's estate, southwest of Mullingar in County Westmeath, Ireland. However, he was not a wealthy man and was careful to economize when he could; he built the first steam boiler for his orchid house himself, as he did the printing press on which he produced the first edition of his orchid manual.[14] Lyons explained that it was published "in the hopes of exciting an interest amongst amateurs, and inducing them to commence their cultivation. They are not difficult to grow, and it is hoped, the following remarks will be found of use to the amateur." Lyons admitted that much of the information he presented already existed in other sources, but he hoped his book had two claims to merit: "first its portable size, and tho' last, not least, the *very very moderate* charge, at which it is presented to the reader."[15] Nobody needed deep pockets or purpose-built shelves to benefit from Lyons' expertise.

The differences between Lyons' and Bateman's orchid books were obvious, yet they had much in common. Both appeared in small runs (although Lyons' book was successful enough to require a second, commercially printed, edition). Like Bateman, Lyons' fascination with orchids had begun with Mexican species and both men had consulted Lindley over the correct botanical nomenclature for the plants they described. (It was Lindley who largely created the modern orchid family, the Orchidaceae.[16]) In both cases, the advice on cultivation was based on attempts to recreate their native conditions; Lyons noted that an orchid collection would include plants from many countries and climates, so they could not all be treated alike. For example, after learning that Trinidad (where many of his plants had come from) had heavy dews, Lyons described his attempt to recreate these conditions:

> I admitted steam every night for some hours amongst the Plants, the atmosphere of the house resembled a London fog, except that it was not so cold; I could not see a yard before me, yet the plants throve wonderfully.

He also repeated Bateman's advice that even tropical orchids need a winter to rest in, berating ignorant amateurs who still assumed *"that all Tropical plants should be kept in constant vegetation, as if they enjoyed an eternal summer."*[17] The orchid grower's goal ought to be "to imitate nature" if they were "to ensure success." That "must be our object and practice":

> Many growers consider that a close, humid, and insufferably hot atmosphere, with constant shade are absolutely necessary toward the growth of Orchideae, nothing in my humble opinion can be more unnatural or more erroneous.[18]

Thanks to the efforts of Lindley, Bateman, and Lyons, orchids would gradually be reprieved from meeting a hot death in Britain's greenhouses. Yet despite the two books' similarities, Lyons' was clearly aimed at a readership very different than Bateman's, who assumed (no doubt realistically) that anyone who could afford his massive book would not be sullying their own hands with orchid compost, and so his guidance focused on the correct instructions to give one's head gardener. By contrast, Lyons expected his readers to be more actively involved, perhaps because he had a generally low opinion of gardeners (he devoted many pages of his book to criticizing them, noting how few followed "the profession as a matter of science," but instead were ignorant and impertinent, spending "their evenings in pothouses [taverns] and their days in cheating their employers"). One of his goals seemed to be helping his readers to avoid having to rely too heavily on such men ("dirty and unshaven, in greasy clothes, foul linen, hats and shoes, like their hair, unacquainted with the luxury of a brush"), so he included much practical guidance.[19] Among Lyons's tips were instructions

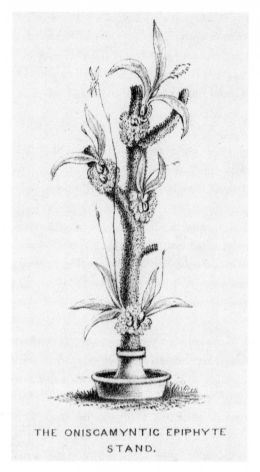

THE ONISCAMYNTIC EPIPHYTE
STAND.

Figure 17. John Charles Lyons succeeded in making orchid growing accessible to
many more people, but had less luck in introducing the word "oniscamyntic" (woodlouse-
repelling) into English. (From *Remarks on the Management of Orchidaceous Plants*, 1843)

for making appropriate stands on which to grow epiphytic orchids, a design of
his own which he proudly named the "oniscamyntic epiphyte stand" (from *onis-
cus*, the wood-louse, and *amuno*, to repel; a word that, unaccountably, has never
really caught on).

The democratization of orchid-growing that began with the publication of
Lyons' book was a feature of the way in which Victorian Britain was chang-

ing more generally. Steam-powered printing presses—combined with cheaper, machine-made paper—were rapidly lowering the prices of books and magazines. By midcentury, literature on every possible subject was in the hands of an ever-increasing readership, thanks in part to organizations like the Society for the Diffusion of Useful Knowledge (SDUK), whose publications—especially the *Penny Magazine*—brought an endless miscellany of fascinating facts to thousands of readers. The Gradgrinds of the SDUK were determined to use the new industrial printing technologies to force facts into the heads of the workers, a commitment that earned it the nickname of "the steam reading society." Through its efforts (and those of rival groups, such as the evangelical Religious Tract Society) many things that had once been the preserve of a small, wealthy elite—from education to the right to vote—would gradually spread through society. Orchids were among the aristocratic delights that gradually made their way into middle-class (and eventually even some working-class) homes.

The year 1851 marked an important change in British society. Millions of people visited London's Crystal Palace to see the Great Exhibition and the newspapers commented—with evident surprise—on how orderly and well-behaved the largely working-class crowds had become, and how comfortably Britain's classes mixed. In the same year, the census revealed that, for the first time in the history of any nation, more Britons lived in towns and cities than lived in the countryside—dramatic evidence of a new, urban society that left many pining for flowers and fresh air. And an 1851 census of church attendance revealed that (despite having been rigged in favor of the established, Anglican church—the Church of England) only one in four of Britain's people actually worshipped in an Anglican church. A similar number attended nonconformist places of worship—such as chapels and meeting houses—but half the country didn't go to church at all. The coincidence of these events and facts led many to wonder what kind of country Britain was becoming: a nation of socially mobile, godless town-dwellers, perhaps, in sharp contrast to the life of the previous century, when God-fearing rural types knew their places and generally stayed in them.[20] As people pondered these changes a gardener called Benjamin S. Williams published his own, modest contribution to the wider changes that were transforming the country. In a series of articles in the *Gardeners' Chronicle* entitled "Orchids for the million" [*sic*] (1851), Williams proclaimed that anyone could and should grow these once-elusive flowers. Orchids helped cement the revolution that gradually led to more and more Britons spending their Sundays in the garden or greenhouse, instead of the church.

"Orchids for the million" was written at the suggestion of Lindley, the paper's editor, and was introduced by Henry Bellenden Ker (writing under the pseudonym Dodman), who explained that Williams was his neighbor's gardener. Ker had first sought Williams' help when he became interested in orchids, initially believing "that the cultivation of these plants is attended with great difficulty . . . and that the secret of good culture is only known to a few." The idea that it took a rare and extraordinarily skillful person to grow these rare and extraordinary flowers persisted well into the twentieth century, which no doubt added to the flower's mystique. Ker was delighted to discover that Williams' system for growing orchids was simple and affordable, so he had urged the gardener to write down and publish his instructions. Ker was an active member of the SDUK, a close friend of its founder, Charles Knight, and a great believer in working-class education, so it is not surprising that he should have encouraged his neighbor's gardener to better himself. Ker criticized Bateman, who "in the preface to his great work, intimates that [the orchid's] culture is to be left to the aristocratic, whilst the more humble florist is to be confined to his Carnations, Auriculas, Dahlias, and such-like flowers."[21] The democratically minded Ker was having none of this — orchids for all!

With Ker's support and encouragement, Williams' series became a successful book, *The Orchid Grower's Manual* (1852), that would eventually go through six editions. Williams was the son of a gardener who worked alongside his father from the age of 14. He taught himself most of what he knew about orchids, eventually becoming a prize-winning grower, author, and proprietor of a successful nursery business. As his son wrote in Williams' obituary:

> His name will be handed down the vistas of time to future generations associated with Orchids — the Royal family of plants; to their study and cultivation the most important part of his active life was devoted with a zeal and enthusiasm that knew no bounds, because his whole soul was wrapt up therein.[22]

A humble gardener becoming an author and businessman, thanks to steam-powered printing and the efforts of groups like the SDUK, exemplified the wider changes in British life; *The Orchid Grower's Manual* was a democratic, almost subversive volume that would help undermine Bateman's vision of an orderly, deferential, floral society. Orchids might be "the Royal family of plants" but they were increasingly to be found in commoners' gardens, greenhouses, and living rooms. Among those who began to take an interest in these increasingly widespread flowers was a middle-aged naturalist, an invalid with a vast expertise in barnacles, Charles Darwin.

5

Orchis Bank

In May 1862, London's *Athenaeum* newspaper reviewed yet another new book on orchids, comparing the prevailing "fashionable fancy for cultivating orchids" to the Dutch Tulip mania of the seventeenth century, during which the price of a single bulb could fetch a price equivalent to several thousand pounds today.[1] The reviewer felt the British fashion was partly due to that fact that orchids "can only be largely and successfully cultivated by the wealthy — more particularly the rarer and grander exotics" — they were luxury items, testimony not merely to the taste but also to the wealth of their possessor. Nevertheless, interest in them had risen to the point where they had "become articles of trade, and even of public auction" where anyone could come along to watch the typical "orchidaceous maniac" pay extraordinary sums to possess a new or rare orchid. However, while the book being reviewed was "a good botanical monograph, with an entomological aspect" on what was obviously a highly fashionable subject, the reviewer had to admit that "popularly it has hardly any place."[2] A slightly surprising verdict given that the book's author was Charles Darwin, already well-known as the author of a successful travel book (his *Journal of Researches*, 1839, which we now call the *Voyage of the Beagle*) and of an extremely controversial work on evolution (*On the Origin of Species by means of natural selection*,

1859). Yet, other reviewers agreed that Mr. Darwin's book on orchids was too technical to sell very well. The *Saturday Review* felt that, despite the "masterly manner" in which the subject was treated, "the nature of its details will somewhat circumscribe its publicity."[3]

Given the extraordinary impact of the *Origin*, we must ask why Darwin's follow-up book was a rather dry and technical tome on orchids rather than a best-selling work on human evolution—a topic he had previously avoided and which most of his readers were probably expecting him to finally address.

Part of the answer lay with the *Origin* itself, which—despite seeming rather long and detailed to modern readers—was regarded by its author as no more than the "abstract" of the full-length book he still planned to write. Darwin had been forced to rush his book into print rather sooner than he'd originally anticipated, because another naturalist, Alfred Russel Wallace, had hit on the idea of natural selection independently. Darwin and Wallace's idea was announced at a meeting of London's Linnean Society in 1858, but made almost no impression on the assembled gentleman of science; in the absence of detailed supporting evidence, natural selection seemed little more than speculation. Wallace had neither the time nor money to write a major book on the issue because, unlike Darwin, he had to earn his living, collecting exotic birds and insects for wealthy, stay-at-home collectors. So, while Wallace was sweating away in the tropics, Darwin used his leisure to produce a shortened version of his planned "big species book" and began the process of persuading the world of the fact of evolution.

The *Origin* was, of course, a sensation when it appeared and while some attacked its speculations as incredible, even heretical, others accepted Darwin's ideas immediately. However, many members of the scientific community, those whose opinion Darwin valued most, simply kept an open mind. Darwin might well be right, but the scientific men wanted more evidence. Darwin's close friend Joseph Hooker, who had been the first person ever to hear of Darwin's theory, had initially responded that, "I shall be delighted to hear *how* you think that this change may have taken place, as no presently conceived opinions satisfy me on the subject."[4] It took Darwin fifteen years of regular letters—mostly about plants—to fully persuade Hooker that natural selection was indeed the key means by which "change may have taken place."[5] The *Origin* won Darwin many more converts, but he knew his work was not yet done and he started mining his notebooks for the necessary supporting evidence.

The popularity of orchids may have prompted Darwin's interest in them, but his son Francis (who was an active participant in his father's botanical research) was convinced that convenience had been his father's main incentive; several species of native British orchids grew wild near the Darwins' home at Downe in Kent, which is one of the richest orchid counties in England.[6] Just half a mile from the house was a spot where Charles and his wife Emma loved to walk, where wild orchids grew in such profusion that they christened it "Orchis Bank."[7] So, as soon as Darwin finished his big books on barnacles and species, his letters and notebooks began to fill with orchids. Originally, they were to have been just one chapter of a book on both animals and plants that would supply much of the detailed evidence omitted from the *Origin*, but—as Darwin explained—the orchid chapter had "become inconveniently large" so he had decided to publish it separately, under the title *On the Various Contrivances by which British and Foreign Orchids are fertilised by insects, and on the good effects of intercrossing*. Despite focusing on one of the most popular groups of flowers, Darwin clearly wasn't trying to write a best-seller (potential purchasers could be forgiven for falling asleep before they'd even gotten through the title). The text inside was equally dry and long-winded, but Darwin felt its length was justified: "Having been blamed for propounding this doctrine [natural selection] without giving ample facts, for which I had not sufficient space in that work, I wish here to show that I have not spoken without having gone into details."[8]

Darwin certainly went "into details" in his orchid book, which contained almost 400 pages of exhaustive descriptions of orchid anatomy and fertilization. These details served more than one purpose; the *Saturday Review* suggested that *Orchids* "will escape the active, and often angry, polemics" that Darwin's previous book had aroused.[9] Darwin may have felt that it was time to take evolution out of the limelight and have it debated in more private, scientific circles, by those whose judgment he respected. The men of science were a very diverse group who included many respected experts we would now classify as amateurs (but virtually no women). They not only needed more evidence, but had to be persuaded that natural selection was more than just a theory—it was also a practical guide to real scientific work.[10] Darwin's detailed explanations of his observations and experiments showed natural selection at work. As another reviewer noted, orchids had been studied by some of the most world's most skillful botanists, yet nobody before Darwin had demonstrated "the function of their various parts. Hence orchids have always remained an unsolved enigma."[11]

Despite the density of the botanical minutiae, *Orchids* was a modest success.

Although they doubted its appeal, reviewers acknowledged its value. The *Parthenon* weekly described it as a "valuable work . . . of especial interest to botanists," but also "addressed to all those minds which have busied themselves with the controversy so intimately connected with Mr. Darwin's name."[12] Another noted that, whether or not one accepted evolution, "there can be but one opinion as to the value of this work as a most important addition to our stock of scientific knowledge."[13] The initial print run of 1,500 copies sold steadily, allowing Darwin to produce a much-expanded second edition in 1877.[14] Despite its almost overwhelming detail, *Orchids* was a key weapon in Darwin's campaign to win over skeptics by showing evolution at work.

Hooker himself reviewed *Orchids*, telling his readers that

> observations, to lead to any good results, must not only be systematically and carefully, but intelligently made; they must, in fact, be suggested by some previous idea, and collected for the support or the contrary, of some possible or probable truth; and the wider the application of that truth, the more fruitful and suggestive will be the accumulated observations directed to its elucidation.

This, for a scientific man, was the whole point of natural selection—it was a "possible or probable truth" that would guide further research. No longer would naturalists roam the countryside collecting beetles or buttercups at random; instead each creature became evidence for or against natural selection, and the centuries-old tradition of natural history would gradually be transformed into something resembling modern biology.

Once we understand why Darwin wrote his orchid book, it becomes much clearer why it is so technical; the details are vital to its long, complex, but ultimately compelling argument about how evolution works and—perhaps even more importantly—a demonstration of how natural selection would transform research for working naturalists. As Hooker argued, what Darwin's orchid book showed was that "all our previous notions were wrong, and most of our observations faulty," and it was thus "a great triumph, that cannot fail to secure to its author a more attentive hearing for his ulterior views [i.e., evolution] than these have hitherto gained."[15] However, even Hooker had to admit "that Mr. Darwin's is no work for the general reader," because of the complexity of both orchid flowers and the analysis of their form and function. To understand the full significance of Darwin's achievement, we first need to understand orchids a little better.

Every Trifling Detail

Darwin admitted that his readers would need "a strong taste for Natural History" if they were to appreciate the book. However, he was confident that he could persuade those who possessed such a taste, that explanations based on scientific laws, such as evolution, were every bit "as interesting" as the explanations offered by those who were convinced that "every trifling detail" of each orchid's structure was "the result of the direct interposition of the Creator."[16] This is typical of Darwin's modest but unabashed tone. He did not claim his view was superior, much less that it was certain to be correct, merely that it was "as interesting" as the idea of special creation (yet at the same time he seemed to gently mock those who could imagine their God being forced to labor over the "trifling details" of every insignificant living thing).

But what were these trifling details?

To appreciate orchids, we need to know a bit of basic botany. Most flowers contain both female and male reproductive organs. The male ones are stamens, usually consisting of slender filaments topped by the anthers that contain the pollen. The female parts are called carpels (or pistils), and consist of an ovary (containing the ova, equivalent to an animal's eggs) above which is a stalk called a style that is surmounted by a stigma. The stigma is the receptive surface where the pollen must first fall before the ova can eventually be fertilized.[17] Once fertilized, the ova swell into seeds and the ovary typically becomes the plant's fruit.

However, orchids are different: their stamens and carpels are not separate, but are fused into a single structure called the column (or, more technically, the gynostemium), which is the main feature used to identify an orchid. At the column's top is either a single stamen, or two or three fused together, which hold the pollen, but instead of being loose grains (as in most plants), orchid pollen grains are typically stuck together in large clumps that, together with their stalks (caudicles), are called pollen masses or pollinia (the singular is pollinium). Below the pollinia is the stigma, which usually takes the form of a shallow, sticky bowl into which the pollinia fits.

Given that the pollinia are only millimeters away from a receptacle that is adapted to receive them, one would assume that orchids are invariably self-fertilized (and some are), but in between the (male) pollen and the (female) stigma is what Darwin considered the most extraordinary feature of orchids: the rostellum, a floral chaperone that both unites and separates the male and female parts. Darwin set out to discover why it was there and how it had been formed.

Darwin was not the first to notice that orchids, like many other flowers, seem structured to *avoid* self-fertilization, despite it being such a readily available option. He had argued in the *Origin* that this was the rule throughout life's kingdoms; nature may resort to self-fertilization from time to time, but avoids it in the long run. Darwin believed that evolution explained why this was so. The key to his theory was variation. Each plant and animal is a little different from its parents, often in seemingly random ways. These variations are not, of course, entirely random, in that plants cannot suddenly sprout wings or feet, but they are random in the sense that there is no predictability about whether each change will enhance or reduce the organism's ability to survive. (The full explanation for this—derived from modern genetics—was not discovered until long after Darwin's death, of course, but his ideas were basically correct.) He noted that some variations would increase an organism's chances in the struggle for existence, perhaps by improving its ability to obtain food, to survive a drought, or to attract a mate. Everywhere around him, Darwin recognized that an intense competition between living things—for food, shelter, or mates—"inevitably follows from the high rate at which all organic beings tend to increase." This competition is not only with other organisms (of both the same and different species), but is also an effort to adapt successfully to their environments (as he wrote in the *Origin*, "a plant on the edge of a desert is said to struggle for life against the drought").[18] Given the intensity of what he called the struggle for existence, any variation that improved an organism's chances of surviving and reproducing (by comparison with its competitors) would be passed on to its descendants, ensuring those descendants became more widespread. (And, of course, any that had negative consequences would tend to be eliminated.) Over the vast, slow ages of geological time, variation, competition, and inheritance combined to create new kinds of animals and plants. This was the process he called "natural selection."

Variability was the key to natural selection and Darwin was convinced that self-fertilization was its enemy; cross-breeding not only tended to produce hardier, healthier plants or animals but also increased their variability. These insights provoked deeply personal reflections for Darwin, who had married his first cousin, Emma, and worried that this decision had produced their rather sickly children, three of whom died very young. If inbreeding was bad for Charles and Emma's offspring, self-fertilization (the ultimate form of inbreeding) ought to be especially bad. He began his botanical researches soon after his marriage to test the idea that inbreeding was harmful—and the flowers seemed to confirm his worst fears.[19] Hence the subtitle of the orchid book: "on the good effects of intercrossing."

Darwin's greenhouse full of orchids was a tool to test his ideas. Not only were his experiments tinged with a faint guilt over his decision to marry his cousin, he fully acknowledged that if it could be proved that a large and successful group of plants, such as orchids, were commonly self-fertilized, it would undermine his evolutionary theory. To some degree, the credibility of natural selection depended on the role of the rostellum since it seemed to be the only thing preventing self-fertilization—and Darwin was convinced that the acceptance of his theory depended on demonstrating that nature avoided self-fertilization.

As Darwin got to work on his orchids, he soon found various allies to help him. As with all his books, *Orchids* is full of acknowledgments to many people; Hooker stood at the head of the list, of course, but Darwin thanked "several gentlemen for their unremitting kindness in sending me fresh specimens," including James Bateman, as well as "my kind assistants," who ranged from Irish vicars to his own children, all of whom had displayed, "the most liberal spirit."[20] And Darwin also got important help from the orchids themselves, as he gradually discovered the properties of a tiny part of the flower's anatomy that nobody had previously fully understood. The pollen masses (pollinia) were attached to a sticky disc that was usually covered by a protective pouch (technically called the bursicula or fovea, but as Darwin chose to avoid these terms, so shall we). The pouch protects the sticky disc from exposure to the air; once it is uncovered, the plant's "glue" quickly hardens. This tiny detail provided the key to a full understanding of orchid fertilization.

Beautiful Contrivances

Most of Darwin's chapters had a similar format: he described each species' structure in detail, with the help of a diagram or two, and then explained how the delicate machinery of the flower promoted cross-fertilization. He commenced with

Orchis mascula, the common British Early Purple orchid, which he probably collected himself from Orchis Bank (but please don't copy him, there or anywhere else: Darwin's site is now a nature reserve; collecting anywhere without the landowner's permission is illegal; and, of course, collecting endangered species is illegal everywhere). At the bottom of the orchid's flower is the enlarged petal known as the lip or labellum (from the Latin *labia*, a lip) and, as Darwin noted, it "forms a good landing-place" for insects, who are attracted by the flower's color and scent, which promise a sip of nectar (although, as we shall see, orchids don't always make good on this promise). To reach the nectar, the insect pushes its head into the flower, brushing against the pollinia at the top of the column; this triggers the protective pouch to rupture, allowing the sticky pollinia to become firmly glued to the insect's head. Darwin encouraged his readers to try imitating the process for themselves, by inserting a sharp pencil into an orchid's flower; it will emerge, as he illustrated, with the pollinia firmly stuck to it.

From both Darwin's and the orchid's perspectives, so far, so good; the pollinia are released, the function of the protective pouch and the sticky disc are clear. But the most remarkable part of the story has yet to occur. The insect (or pencil) emerges with the pollinia, "firmly cemented to the object, projecting up like horns," but if the insect were to crawl into the next flower with the pollinia in this position, they would simply brush off onto the flower's male pollinia. Unless the pollinia land on the female stigma, pollination cannot occur. However, as Darwin watched the pollinia on the tip of a pencil or insect's head, he saw another remarkable phenomenon. As the plant's "glue" dried, it contracted, causing the slender threads (the caudicles) that attach the pollinia to the disc to bend, which pushed the pollinia forward. The bending resulted in the insect's pollen payload pointing forward (instead of up) — in exactly the right position to miss the next flower's pollinia and touch the female stigma instead. And the brief delay between the pollinia attaching and the forward bending (which Darwin christened "depression"), was long enough to allow the insect to move to a new plant, making it more likely that each flower would be fertilized with pollen from a different plant. The "good effects of intercrossing" were secured by a combination of the orchid's anatomy and the insect's behavior.

For this extraordinary mechanism to work, everything has to be just right: the positions of the pollen and stigma, the stickiness of the glue (as Darwin noted, "the firmness of the attachment of the cement is very necessary, for if

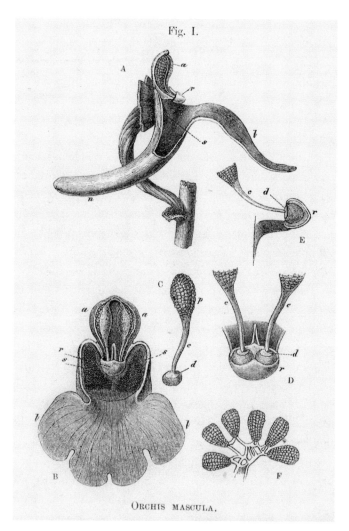

Fig. I.

ORCHIS MASCULA.

Figure 18. Using diagrams and explanations, Darwin showed how the flowers of orchids were elaborate mechanisms for enlisting insects into ensuring cross-fertilization. (From *On the Various Contrivances by Which British and Foreign Orchids Are Fertilised by Insects, and on the Good Effects of Intercrossing,* 1862)

the pollinia were to fall sideways or backwards they could never pollinate the flower"); the speed at which the glue dried; and the degree to which it contracted, bending the caudicles as it did.[21] Above all, the insect and the flower had to fit one another; many orchids could not be pollinated by an insect that was either too large or too small. No wonder Darwin argued that "the contrivances by which Orchids are fertilised, are as varied and almost as perfect as any of the most beautiful adaptations in the animal kingdom."[22]

However, such beautiful perfection raised an inevitable question for Darwin and his readers; surely, such "contrivances" revealed the hand of God at work? As the reverend William Paley (whose works Darwin read and loved as an undergraduate) had famously written, "There cannot be design without a designer; contrivance without a contriver."[23] This argument, often known as the argument from design, was a cornerstone of the English tradition of natural theology, which argued that the perfection of organisms like orchids proved that a creative intelligence, a divine creator, must be responsible for their existence. Moreover, the tight fit between orchid and insect, each supplying the other's needs (for fertilization or nectar), proved that the Creator was benevolent. And the extravagant and gratuitous beauty of flowers such as orchids, which so delighted their human aficionados, proved beyond doubt that the Creator was none other than the God of Christianity, deeply concerned with the edification and happiness of his creation, especially for humanity, his masterpiece.[24]

Some of those who reviewed Darwin's orchid book seem to have been confused by his repeated use of the word "contrivance," and went so far as to describe *Orchids* as a work of natural theology. The *Parthenon*'s reviewer commented that Darwin's researches had clearly not shaken "for a moment his allegiance to the great Creator of the universal whole."[25] Another asserted that, regardless of Darwin's views, the book would be equally interesting to "the believer in the doctrine of secondary causes [as] to him who takes the older and more orthodox view of Paley and his followers." The review concluded:

> From [Darwin's] armoury of facts the former will draw many new weapons wherewith to assail the older belief; and the latter, securely entrenched behind his impregnable rampart of faith, may find in its pages new and marvellous instances of design, by the aid of which he may seek to repel the ardent assailants.[26]

The possibility that a supporter of the design argument "may" find new supporting examples in the orchid book seems to have concerned one reviewer so much that he even criticized Darwin for his "unhesitating teleology":

> Mr. Darwin asserts that the Orchid secretes nectar *in order* to attract insects. Surely it is enough for the philosopher to note that the nectar is secreted, and the insects attracted, without perilously undertaking to assert that the nectar is secreted specially for that purpose, when the secretion may have many and more important parts to play.[27]

One of the book's most distinguished reviewers was George Campbell (8th Duke of Argyll), a staunch promoter of natural theology who opposed Darwin's ideas courteously but emphatically. His review commented that despite Darwin's professed intention of avoiding supernatural explanations, Darwin seemed to embrace the language of natural theology. As Argyll commented, in the orchid book the words "contrivance," "curious contrivance," and even "beautiful contrivance" were "expressions which recur over and over again." He therefore asserted that "intention is the one thing which [Darwin] does see, and which, when he does not see, he seeks for diligently until he finds it." Argyll drove home his point by noting that Darwin often used human artifacts, such as spring traps, as metaphors for the parts of orchids.[28] As Argyll and his readers knew very well, this kind of analogy was at the heart of Paley's book *Natural Theology* (1802), in which he famously wrote that anyone seeing a watch for the first time would be forced to conclude that it had been made by an intelligent watchmaker (even if they'd never seen such an artisan at work), since a preconceived *purpose* had clearly determined its design.[29]

In a letter to his friend, the Harvard botanist Asa Gray, Darwin ruefully acknowledged that Argyll's review was "clever," yet he still could "not see that it really removes any of the difficulties of Theology."[30] Yet when Gray himself, who remained a lifelong Christian despite his support for Darwin, reviewed *Orchids*, he chose to emphasize many of the same things that Argyll had done.[31] He even went so far as to say that if Darwin had published *Orchids* first, he might "perhaps" have found himself canonized rather than anathematized, "even by many of those whom his treatise on the origin of species so seriously alarmed," because the orchid book "would have been a treasury of new illustrations for the natural theologian." Although Gray had to admit that, to achieve this goal, Darwin would have had to keep back "a few theoretical inferences."[32]

In a later article, Argyll recounted actually meeting Darwin and discussing the orchid book with him:

> I said it was impossible to look at [orchids] without seeing that they were the effect and the expression of Mind. I shall never forget Mr. Darwin's answer. He looked at me very hard and said, "Well, that often comes over me with over-whelming force; but at other times," and he shook his head vaguely, "it seems to go away."[33]

Why, for Darwin, did the argument from design "go away"? Why didn't the orchid's beautiful contrivances convince him, as they did the Duke and Gray, that there must be a designing mind behind the natural world? And, if Darwin really didn't believe there was conscious design in nature, why did he sometimes write in a way that seemed to imply he did?

Darwin is sometimes presented as a militant atheist, determined to destroy people's faith in God, but there is no evidence to support this view. He once wrote to Gray that the whole question of God was "too profound for the human intellect. A dog might as well speculate on the mind of Newton.–Let each man hope & believe what he can."[34] In his published writings, Darwin was usually careful to avoid religious controversies and often expressed himself with an ambiguity that allowed his readers to "hope and believe" whatever they wished. However, Darwin — like an increasing number of Victorian scientific men — was firmly convinced that God had no place in *scientific* explanations. Indeed, excluding God from one's explanations was gradually becoming part of the definition of what it meant to be scientific in nineteenth-century Britain.[35] So, regardless of what men like Argyll or Gray chose to hope and believe, Darwin knew that his fellow men of science would need a thoroughly scientific explanation of the orchids' beautiful contrivances if they were to be persuaded that these were the products of blind, chance-driven natural selection. One of the key questions that he asked of his orchids was this: how can there be contrivance *without* a contriver?

Like other reviewers of *Orchids*, Gray chose to emphasize the extent to which it resembled traditional popular natural theological books, describing it as a new addition to that class of books that are "perhaps the most attractive to old and young," those "which describe the habits and doings of insects." He made it clear that Darwin was trying to explain the adaptations of orchids by reference to "secondary laws," not to the direct action of a Creator, but he nevertheless argued that the facts and observations Darwin described would be equally fascinating "irrespective of all theories of origination, and perhaps as readily harmonized with old views as with new." Gray probably assumed that this was the best strategy for persuading the more devout among his fellow Americans to actually read *Orchids*, and read it with an open mind (he gave the evolutionary argument but added "we have no desire nor particular occasion to reopen this question now").[36] By contrast with his cautious public comments, in a private letter Gray congratulated Darwin on the way in which "your beautiful flank-movement with the Orchid-book" was gradually winning over naturalists who had been opposed to the *Origin of Species* just three years earlier.[37] A delighted Darwin wrote back, "Of all the carpenters for knocking the right nail on the head, you are the very best: no one else has perceived that my chief interest in my orchid book, has been that it was a 'flank movement' on the enemy."[38] Not only was Darwin not supporting natural theology, he was trying to turn the design argument against its own supporters.

Darwin's argument was a subtle one, but it had a couple of key aspects that make its appeal clear. Firstly, he took advantage of the widespread interest in orchids by inviting the skeptical to see for themselves. For example, in describing the shape of a particular orchid's nectary (the part of the flower that holds the nectar), Darwin noted that natural selection had produced vertical ridges along its sides, so that the long, flexible proboscis (the organ through which a butterfly or moth sucks the nectar) was guided into the nectary's depths without bending. If that seemed implausible to any of his readers, Darwin suggested that "he who will insert a fine and flexible bristle into the expanded mouth of the flower between the sloping ridges on the labellum, will not doubt that they serve as guides and effectually prevent the bristle or proboscis from being inserted obliquely into the nectary." In other words, if you don't believe me, try it yourself. This kind of argument appealed to everyone from working naturalists to the wider community of orchid enthusiasts, whether they were commercial nurserymen or amateur growers. Even the *Athenaeum*'s reviewer (who was generally skeptical about natural selection) acknowledged that the book had clearly demonstrated that flowers were adapted so as to make them dependent on their insects, and urged anyone who doubted this: "Try Mr. Darwin's experiment, and you will arrive at his conclusion."[39] Almost anyone could test Darwin's claims for himself and, by doing so, join the scientific world; Britain's legions of self-taught gardeners were not merely consumers of scientific writing, but could be active participants in scientific discovery (Darwin wrote regularly to magazines like the *Gardeners' Chronicle* and *Cottage Gardener* seeking information on various points). Darwin's theories suggested new experiments and observations, whereas the claim that God designed the orchid did not suggest experiments or new ideas to test, since one's belief rested on faith, not evidence.

The idea of being able to participate in science may have caught the attention of orchid growers, perhaps even flattered them a little, but it would not have persuaded them that Darwin was right. He had to produce evidence to support his main contention that there could indeed be contrivance without a contriver. He built up to this argument through his long, detailed discussion of numerous orchid species, in which he explained ("perhaps in too much detail," he admitted) why there was such a close fit between insect and orchid, and how it ensured cross-pollination in each species.[40] In Catasetums, for example, the pollinia are ejected with such force that they not only stick firmly to the insect, but send it flying off to another plant straight away, thus avoiding the danger of self-pollination. In the Brazilian bat orchid, *Coryanthes speciosa*, the labellum is bucket-shaped and fills with a sugary secretion. Bees that fall in pollinate the flower as they crawl out. Darwin noted that if the bucket were dry, the bees could fly away, but once their wings are wet, they cannot fly and have to crawl through a narrow exit where they necessarily brush against the stigma and pollinate the flower.[41] Yet, despite his delight in the near-perfect complexity of such arrangements, Darwin was well aware that perfection might be the result of divine design; only *imperfection* proved that natural selection had been at work.

Among the most baffling orchids Darwin considered was one discovered by the explorer Richard Schomburgk (who is also credited, slightly inaccurately, with discovering the giant Amazonian water lily, *Victoria regia*, now *Victoria amazonica*, whose meter-wide lily pads fascinated Victorians). The orchid Schomburgk found possessed three completely different flowers growing on a single stem. Even more remarkably, the flowers were not merely those of different species, but came from different genera: *Catasetum tridentatum*, *Monachanthus viridis*, and *Myanthus barbatus*. As Lindley wrote at the time, "such cases shake to the foundation all our ideas of the stability of genera and species."[42] Darwin analyzed Schomburgk's mystery plant in detail and showed that the

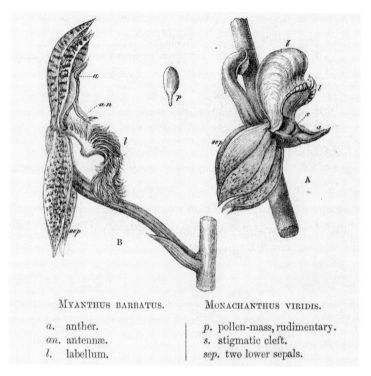

MYANTHUS BARBATUS.

a. anther.
an. antennæ.
l. labellum.

p. pollen-mass, rudimentary.
s. stigmatic cleft.
sep. two lower sepals.

MONACHANTHUS VIRIDIS.

Figure 19. The orchid that shook "all our ideas of the stability of genera and species."
The two species that Darwin identified as being sexual forms of *Catasetum tridentatum*.
(From *On the Various Contrivances by Which British and Foreign Orchids Are
Fertilised by Insects, and on the Good Effects of Intercrossing*, 1862)

three "species" were nothing of the sort, merely three forms of the same species: the flower to which the name *Catasetum tridentatum* had been given was in fact the male form of the flower (possessing a stamen but no stigmas), while the supposed *Monachanthus* was a female form (stigma but no stamen), and *Myanthus barbatus* was the hermaphrodite form (stamen and stigma). The three forms were all *Catasetum tridentatum* but looked so different that when botanists had first seen them in isolation each had been given a different name (not merely misidentifying them as different species, but placing them in separate genera). The mistake was perhaps understandable, given that this was the first orchid to be identified with its male and female flowers on male and female plants (dioecious is the technical term for this kind of plant, while plants with male and female flowers on the same plant are said to be monecious).[43] The case fascinated Darwin, partly because it provided suggestive hints as to how separate sexes might have evolved (as they have in many plant species), but also because the male and female flowers contained useless rudiments of the missing organs. As Darwin wrote, "Every single detail of structure which characterises the male pollen-masses is represented, with some parts exaggerated and some parts slightly modified, by the mere rudiments in the female plant." Similar evidence is found throughout the natural world (for example, whales have the vestiges of legs, because they evolved from land-dwelling animals that went back to the water); a divine creator would have to be frankly eccentric to design his creatures with the useless remnants of organs they had never needed and which suggested an evolutionary history that had never in fact occurred. Darwin concluded that it would not be long before naturalists would learn with "surprise, perhaps with derision" that

> grave and learned men formerly maintained that such useless organs were not remnants retained by inheritance, but were specially created and arranged in their proper places like dishes on a table (this is the simile of a distinguished botanist) by an Omnipotent hand "to complete the scheme of nature."[44]

By contrast, Darwin's analyses demonstrated that the varied and extraordinary forms of orchids could best be understood as variations on a single ancient form, of which he provided a simple diagram.

A flower is made up of successive rings (which botanists call whorls) of floral parts. The outermost whorl usually consists of the green sepals that form the calyx (which protects the flower's bud before it opens), inside this is the corolla, made up of petals, then come the stamens with the carpels inside them. Some flowers consist simply of four whorls of floral parts (calyx, corolla, stamens, and carpels). Monocots (plants with a single embryonic seed leaf) have their floral parts in threes or multiples of three. The ancestral orchid probably looked like the classic form of a simple monocot, such as a *Tradescantia* flower (also known as Spiderwort or Wandering Jew); a symmetrical flower with three sepals, three petals, six stamens and three carpels. However, in orchids, evolution had transformed each whorl, modifying or fusing its parts so that, for example, the lowermost of the three petals in most orchids is enlarged to form the labellum, most of the stamens have been lost or become infertile, two of the stigmas have fused to form the column, while the third has become sterile and has been modified to form the extraordinary rostellum. As Darwin concluded, "What an amount of modification, cohesion, abortion, and change of function do we here see!" Which raised an obvious question:

> Can we feel satisfied by saying that each Orchid was created, exactly as we now see it, on a certain "ideal type;" that the omnipotent Creator, having fixed on one plan for the whole Order, did not depart from this plan; that he, therefore, made the same organ to perform diverse functions—often of trifling importance compared with their proper function—converted other organs into mere purposeless rudiments, and arranged all as if they had to stand separate, and then made them cohere?

Darwin found this notion improbable, suggesting instead that "a more simple and intelligible view" was that all the orchids were descended from a common ancestor, "and that the now wonderfully changed structure of the flower is due to a long course of slow modification" (a claim that has now been confirmed by DNA studies).[45]

What the orchids revealed most clearly to Darwin was a lack of foresight that he gradually came to feel was incompatible with any notion of intelligent creation. Consider, for example, the bog-orchid *Malaxis paludosa* (now *Hammarbya paludosa*). In most orchids, the labellum is at the bottom of the flower, so that insects land under the column and its pollinia, but if you examine the developing embryo of an orchid, you will see that the labellum starts off as the upper petal and twists through 180° during the flower's development. That already suggests a somewhat eccentric creator, but in *Malaxis paludosa*, the labellum appears at the top of the flower, having twisted through a full 360° during the flower's development. Surely even a moderately intelligent designer would have simply left it alone? This kind of evidence was discussed by Darwin in letters with Asa Gray, who (like many of his contemporaries) reconciled his Christianity with his evolutionism by presuming that Darwin was right—individual organisms were indeed the products of nature's laws, most notably natural selection—but those laws were God's handiwork. (Just as our planet's seasons were a consequence of the impersonal, mathematically precise law of gravity moving the Earth around the sun: God did not have to get the sun out of bed every morning and propel it across the heavens himself, instead he had created gravity to keep the heavens moving and bring us spring each year.) Taking a theistic view of evolution was what had allowed Gray to give *Orchids* such a favorable review, minimizing the differences between Darwin's view and those of the natural theologians (as Gray himself put it, in the title of one of his articles "Natural Selection Not Inconsistent with Natural Theology").[46] However, Gray seems not to have fully grasped Darwin's argument (or perhaps he resisted its full implications). For Darwin, a case like *Malaxis paludosa* confirmed a complete absence of divine involvement; why make an orchid's labellum rotate through a full circle, when it could simply have been left in its original, embryonic position? The fact that natural selection so clearly lacked foresight was not a problem for Darwin, since he was sure that variations were not directly produced by the organism's circumstances or needs, but were genuinely random—in the sense that they had no preferred direction. However, the lack of foresight made it all but impossible to see how evolution could be considered a divinely ordained process, one intended to perfect or otherwise benefit the organisms it produced.[47]

In the very same letter in which Darwin had congratulated Gray for realizing the orchid book was a "flank movement," he added a postscript, "I sh[oul]ᵈ like to hear what you think about what I say in last Ch[apter]. of Orchid Book on the meaning & cause of the endless diversity of means for same general purpose.—It bears on design—that endless question."[48] How, in other words, did Gray explain the *imperfections* of orchids, apparently clear evidence of their creator's inability to see ahead? Gray ducked the issue for several letters, but was finally forced to acknowledge that "here lives, I suppose the difference between us. When you bring me up to this point, I feel the *cold chill*"—the chilling suspicion that Darwin's universe might indeed be a godless one.[49] Darwin replied that he had "suspected what you would say," but added that "what, I think, ought to give you the severest 'cold chill' is the case of Pouter, Fantail-pigeons &c." These Victorian fancy pigeon breeds were his favorite example in the *Origin*, where he had shown his readers how radically human breeders had changed the common rock pigeon in the comparatively brief period of recorded human history and then asked them to imagine what natural selection could accomplish given the much vaster scope of geological time. He pressed Gray on the point: "were not these variations accidental as far as the purpose man has put them to?"[50] In other words, surely Gray did not imagine God had created the pigeon's variations in order to satisfy the arbitrary whims of pigeon breeders? And if these variations were not divinely planned, why should we assume any of nature's other variations were in any way preordained?

Darwin explained all the modifications of the orchid's form, including the 360° twist in the labellum of *Malaxis paludosa*, as the result of nature only ever being able to select from whatever variations happened to occur. If some change, perhaps to the nature or habits of the orchid's pollinating insect, made it advantageous for the labellum to return to its ancestral position at the top of the flower, natural selection could only work on the available variations and if none of them happened to be "untwisting" variations, the orchid simply went on twisting until it came full circle. Not only could evolution not plan ahead, each variation could only have been preserved because it was immediately "useful to the plant."[51] And what was that use? Most of the complex modifications were there to assist reproduction — and the reproduction of a species is, from an evolutionary perspective, the highest possible good. Ultimately, in Darwin's world, nothing else really matters. Of course there are many ways for plants to get their pollen to other plants; the grasses, which are also monocots, have evidently been massively successful in spreading themselves across the planet, yet their flowers are tiny and insignificant because they rely on the wind, not insects, to spread their pollen. The plants have gradually diversified into numerous families with different strategies for surviving, as a result of different variations having occurred among their ancestors; the grasses and orchids had different sets of variations for natural selection to work on, and their subsequent evolutionary success is entirely explicable, but the original variations occurred purely by chance.

The complex and varied contrivances of orchids make no sense, Darwin wrote, "unless we bear in mind the good effects which have been proved to follow in most cases from cross-fertilisation." (He had spent years in his greenhouse, comparing the fertility and hardiness of cross- and self-pollinated flowers to prove these "good effects.") Among the orchids, natural selection had favored ever-closer links between specific insects and particular flowers, so much so that in a few cases, only one species of insect can pollinate a particular species of orchid. For any species to become so dependent on another might seem dangerous; what happens if your only pollinator (or major food source) becomes rare, or even extinct? That can, of course, happen (many orchids and insects are currently at risk for precisely this reason), but natural selection has no foresight. It might seem paradoxical, but in the short term any accidental modification of the orchid that *reduced* the number of insect species that could pollinate might be *beneficial* to both plant and insect. Why? Because if the pollinator can feed only from flowers that are similar to one another, less of the orchid's precious

From a photograph by W. Ellis.

ANGRÆCUM SESQUIPEDALE AND NATIVE FERNS.

Figure 20. A European collector in search of new orchids, such as the star orchid,
Angraecum sesquipedale. (From William Ellis, *Three Visits to Madagascar*, 1858)

pollen would be wasted on completely unrelated species that it could not fer-
tilize. In exactly the same way, any slight variation in an insect that allowed it
to feed more effectively from a few particular flowers would be to the insect's
benefit (there would be less competition for that food source). Gradually, over
innumerable generations, the fit of orchid to insect would become tighter until,
in some cases, they matched like lock and key. A high-risk strategy in the long
term, perhaps, but (just to repeat, perhaps in too much detail, the main point)
natural selection has no foresight—in the short term the insect had no competitors for the nectar, while the orchid reduced pollen wastage.

Perhaps the simplest but most effective proof of Darwin's argument was
an orchid from Madagascar called *Angraecum sesquipedale,* known by various

other names, including the Star or Comet Orchid. Darwin's specimens of these "large six-rayed flowers, like stars formed of snow-white wax" came from none other than James Bateman himself in January 1862. The beautiful blooms made this a very desirable hothouse flower, but Darwin's attention was caught by the flower's "green, whip-like nectary of astonishing length [which] hangs down beneath the labellum."[52] These extraordinary flowers had first been discovered by a Frenchman, Louis-Marie Aubert-Aubert du Petit-Thouars, who gave the species its name because of those long nectaries (*sesquipedale* literally means "one and a half feet" long, a slight but pardonable exaggeration). The flower and its name first appeared in du Petit-Thouars beautiful book *Histoire Particulière des Plantes Orchidées Recueillies sur les Trois Isles Australes d'Afrique* (1822), of which only two copies survive—one at Kew. He had not tried to import any into Europe for cultivation and it would be 35 years before anyone succeeded in getting the flowers to grow outside their native Madagascar. In 1857, the Reverend William Ellis finally brought some to Britain, but they remained very rare. During the early 1860s, a single plant cost over £20 (the equivalent of over £12,800/$20,250 today); even the wealthy Bateman must have respected Darwin highly to send him more than one.[53]

Yet even as Darwin reveled in his new gift, he was puzzled by the flower's nectaries: they grew to almost 11.5 inches (28 cm) long, but he noted that "only the lower inch and a half filled with very sweet nectar."[54] As he wrote to Joseph Hooker when he received Bateman's gift, "I have just received such a Box full from Mr Bateman with the astounding *Angræcum sesquipedalia* [*sic*] with a nectary a foot long—Good Heavens what insect can suck it"?[55] Just five days later, he wrote to Hooker again, mentioning the same orchid and adding, "What a proboscis the moth that sucks it, must have! It is a very pretty case."[56] Similarly shaped and colored flowers were pollinated by moths, with a proboscis long enough to reach into the orchid's nectary, so the idea of a moth like an insectile elephant, with a snout long enough to reach down and drink the distant nectar, had obviously popped into Darwin's head immediately.

Darwin's orchid book was nearing completion (it would be published in May 1862), but in the three months after Bateman's extravagant gift had arrived, Darwin became convinced that somewhere in Madagascar "there must be moths with proboscides capable of extension to a length of between ten and eleven inches!"[57] He explained exactly how he imagined the moth and orchid must have evolved to fit one another, arguing that if either the moth or flower "were to become extinct in Madagascar" the other would perhaps also become extinct. Nevertheless, he was convinced that once the orchid and its moth were viewed from an evolutionary perspective, naturalists could "partially understand how the astonishing length of the nectary may have been acquired by successive modifications." Darwin had no moth to observe, so he tried his usual technique of inserting various hairs and bristles into the flowers to see if they emerged with the pollinia attached. He discovered that only a fairly substantial "cylindrical rod one-tenth of an inch" (2.5 mm) in diameter could withdraw the pollinia; clearly, only a large moth would succeed. There was variation in the length and thickness of all moths' proboscides, so at some point earlier in Madagascar's history those moths with the longest and thickest noses would have been the most successful at pollinating those orchids that (again, simply because of undirected variation) had the longest nectaries. The long-nosed moths would get best access to the nectar in those orchids that had the deepest nectaries and if the moths varied randomly so that some began to prefer the flowers with the longest nectaries, that preference would be preserved by natural selection and passed on to their descendants. As the moths started to specialize, the pollen from the orchids with the longest nectaries would be increasingly likely to land only on flowers with similar structures. And so the long-nectary variation would be passed on. (Today, of course, we know it's the gene for a long nectary that becomes more frequent, but nobody in Darwin's day understood the mechanism of inheritance, a fact that makes it even more remarkable that he got so many of the details of evolution right.) As a result of this successful pollination, Darwin wrote, "these plants would yield most seed, and the seedlings would generally inherit longer nectaries; and so it would be in successive generations of the plant and moth." There had, he argued "been a race in gaining length between the nectary of the Angraecum and the proboscis of certain moths"; only the largest moths could drink from the deepest nectaries and so each species increasingly specialized in—and eventually came to depend upon—the other.[58]

Nobody had seen a moth with the prodigiously long nose that Darwin predicted, and in the second edition of *Orchids*, he claimed that entomologists had ridiculed him for such a preposterous suggestion. But he remained convinced that nothing else could explain the orchid's structure. Forty-one years after Darwin made his prediction (and twenty years after his death), Darwin's moth was found and given the scientific name *Xanthopan morganii praedicta* (i.e., "foretold") in honor of natural selection having accurately forecast its existence. Even so, while the moth had the right size proboscis, it would not be until 1992 that scientists finally observed (and filmed) it visiting *Angraecum sesquipedale*.[59] And, further evidence for Darwin's claim has been revealed by the fact that the orchid genus *Angraecum* and the sphinx moths that pollinate them are both amazingly diverse in Madagascar; specialization works.[60]

The Star Orchid and its moth were typical of the kind of science that natural selection would make possible. Ever since Theophrastus, natural history had mainly been a science of collecting and observing, watching and describing, naming and classifying. By contrast, the more prestigious sciences, like physics and astronomy, had used their mathematically precise laws to *predict* phenomena, such as the return of a comet or the existence of a previously unknown planet. Thanks to Darwin's orchids, naturalists could now begin to do the same, and natural history had taken a small but crucial step on the road to becoming one of today's most prestigious sciences, modern biology.

Nevertheless, many questions remained unanswered. Why did orchids so often resemble insects, for example? After the orchid book appeared, Darwin's contemporary, the naturalist St George Jackson Mivart, wrote to him to ask "if you would kindly tell me by what action you think the curious resemblance of the Bee, Spider & Fly Ophrys to the several insects has been produced," adding "It is I think far too marked & striking to be accidental!" Mivart, who converted to Catholicism a few years earlier, clearly thought he had detected God's hand in some way. Darwin could offer no explanation, other than to say that he thought the resemblance was probably accidental and in any case exaggerated ("the resemblance is fanciful; the flowers are odd looking &insects are the most natural standard of comparison").[61] That did not deter Mivart from using these orchids as one of many examples of things natural selection could not explain in his anti-Darwinian book *On the Genesis of Species* (1871).[62]

While Darwin acknowledged that he could shed no real light on these curious examples of apparent resemblance, he included an enigmatic footnote in his book that mentioned a fellow naturalist who had "frequently witnessed attacks made upon the Bee Orchis by a bee," upon which Darwin commented, "What this sentence means I cannot conjecture."[63] The solution to these mysteries would not be worked out until the twentieth century, but in the meantime — equipped with the power to view nature not through the rose-tinted spectacles

of natural theology but through the sharply focused lens of natural selection—
biologists would be able to imagine and conduct a vast number of experiments
to test their hypotheses; and whether their guesses were confirmed or disproved,
each experiment suggested others that could be done. No wonder that Darwin
himself, in a private autobiographical note, described natural selection as being,
above all, "a theory by which to *work*."[64]

Plate 6. In their eagerness to announce new orchids, Victorian collectors regularly coined new names, as for this plant, which was called *Phalaenopsis grandiflora aurea.* Many are no longer considered species; this one is identical to *Phalaenopsis amabilis.* (From F. Sander and Co., *Reichenbachia: Orchids Illustrated and Described,* 1888)

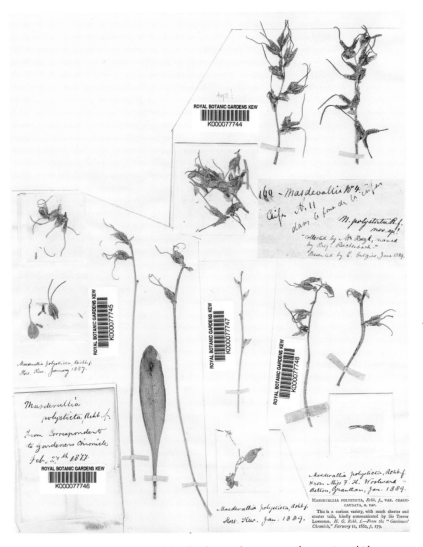

Plate 7. One of the original dried and pressed specimens of an exotic orchid
(*Masdevallia polysticta*) collected by Benedict Roezl in Peru, ca.1877.
The sheet is now in the Herbarium at the Royal Botanic Gardens, Kew.

Plate 8. The elusive Military Orchid (*Orchis militaris*), illustrated by Gavin Bone. (From Jocelyn Brooke, *The Wild Orchids of Britain*, 1950)

Himantoglossum
Hirinum
Lizard Orchid
(Kent. June 1924.)

Plate 22

Plate 9. The Lizard Orchid (Himanto-glossum hircinum), "that legendary flower" that is finally found amid a company of Italian soldiers in Brooke's autobiographical novel, illustrated by Gavin Bone. (From Jocelyn Brooke, The Wild Orchids of Britain, 1950)

Plate 10. The slap heard around the world: Mr. Tibbs (Sidney Poitier) refusing to turn the other cheek to Eric Endicott (Larry Gates). (From In the Heat of the Night, directed by Norman Jewison, 1967)

The Scramble for Orchids

Alongside the humble British orchids that came from Orchis Bank, Darwin's greenhouse contained species brought from all over the world. He made regular use of Joseph Hooker, who happily shared Kew's green riches, but—like any other Victorian gardener—Darwin relied on commercial nurserymen. When he wrote to Hooker about filling a new hothouse (which would be warm enough to house tropical plants), Darwin mentioned that he had been "looking with much pleasure at catalogues to see what plants to get." He was interested in carnivorous plants such as the pitcher plant, *Sarracenia*, together with those that exhibited unusual movements, such as the sensitive plant, *Mimosa*, "& all such funny things." He planned to "beg" Kew for the loan of some orchids, "but I must get priced catalogue."[1]

Among the many tropical orchids Darwin was curious about were the central American genus of gorgeous epiphytes called *Catasetum*. He asked Hooker for some, but apparently Kew had none to spare, since Darwin told him "I have written to Veitch, (whose name you mentioned with Parker & Williams) for 4 kinds; if he fails, I will let you know as I wish intensely to see Catasetum."[2] The "Williams" mentioned was none other than Benjamin Williams, the self-made orchid expert (see chapter 4) whose articles had now blossomed into a successful business, the

Victoria and Paradise Nurseries, of which Darwin was one of many customers. Darwin also got many different orchids from James Veitch and his son (another James), who ran the Royal Exotic Nurseries in Chelsea, central London (when Darwin published his orchid book, he thanked both Parker and Williams and James Veitch junior, who "has generously given me many beautiful Orchids").[3]

The nurseries that supplied Darwin (and many others) with orchids were, of course, a product of the popularization of the plants that Williams had done so much to launch. However, the orchidmania of the 1840s and '50s was nothing compared to its second wave, in the last few decades of the century, which saw more orchids arrive, more orchid-growers buying them, and more profits for the nurseries. London was the heart of this global trade, where orchidmania seems to have been at its most intense, but the passion and the trade were to spread across Europe, to the United States, and beyond towards the end of the century. Part of the attraction, as one Victorian orchid grower's manual explained, was simply that they were long-lasting flowers; an orchid "may be worn even for a long evening, and be as fresh at the close as when newly gathered." And it assured its readers that there was "always a sound and hearty reality about them":

> An orchid-flower means what it says. It does not fall to pieces like a lily; there is no shedding of petals; no dropping away from the peduncle; no self-decapitation, like that of a fuchsia; no collapsing and dissolving, like a spider-wort; — no, there is never any of this; the orchid-flower is neither superficial, nor fugitive, nor insincere. . . . If we mistake not, orchid-flowers have a grand future before them.[4]

One man in particular, Frederick Sander, was so convinced about the future for orchids that he devoted his life to them, eventually dominating the trade to such an extent that he became known as the "Orchid King."[5] He was born in Bremen (in what was then the North German Confederation) in 1847. He worked for a couple of different nurserymen in Germany as a teenager and migrated to England in 1865, supposedly with half a crown in his pocket and speaking little English. He was a short man (5 feet 4 inches tall), but very energetic. He worked for Messrs. Ball's nursery and then for Carters, of Forest Hill in southeast London, where he met an extraordinary orchid collector called Benedict Roezl, who introduced Sander to the flowers that were to obsess him for the rest of his life.

Roezl's early life is mainly a mystery (as it was in his lifetime), but he was apparently the son of a Czech gardener and seems to have been born in Horomeric, northwest of Prague (now in the Czech Republic), in 1824. He began life as a farmer but took to plant collecting after losing his right arm while demonstrating a new piece of agricultural machinery he had invented; the iron hook that replaced his right hand made him an unforgettable, rather piratical figure, which may help explain the aura of larger-than-life legend that came to surround him. According to Roezl's biographer, the Victorian journalist Frederick Boyle (who never turned down the chance for a little myth-making), Roezl was tall and well-built, and never carried firearms but relied on his force of personality to protect him. "As for the wild Indians," Boyle writes, "I fancy that they were over-awed by his imposing appearance; and especially by an iron hook which occupied the place of his left hand, smashed by an accident" (the fact that Boyle has robbed Roezl of the wrong hand is typical of his rather cavalier attitude to facts). In Boyle's view (in a chapter with the refreshingly honest title "The legend of Roezl"), the Czech was "incomparably the greatest of those able and energetic men who have roamed the savage world in search of new plants for our pleasure and enjoyment" and Boyle is quite prepared to believe the claim that Roezl single-handedly discovered 800 new species of orchid.[6] On one occasion he is even supposed to have quelled a hostile black bear "with the power of his eye," and then calmly dispatched 3,500 orchids to London.[7] All in a day's work for this remarkable character. Unsurprisingly, Frederick Sander was impressed with Roezl and rapidly came to share his love of orchids.

Not long after meeting Roezl, Sander met and fell in love with Elizabeth Fearnley, the daughter of a wealthy neighboring landowner, who not only didn't oppose the marriage but settled a considerable sum on Elizabeth. So, in 1876 Sander was able to use his wife's money to buy an established seed business in St. Albans, about twenty miles northeast of London. Sander was originally a general seed merchant (his notepaper described him as "a special grower of peas, cabbage, turnip, mangel and cauliflower"—with no mention of orchids). However, by the 1880s he realized that—thanks to the efforts of men like Lyons and Williams—orchids could be a profitable business; he sent his first collector to South America and was soon growing and selling numerous species in St. Albans. In 1885, Sander changed his firm's letterhead to read "Orchid grower" and within a few years, he was so well known that a collector in the tropics once sent a letter addressed simply to "Orchid King, England" and it reached him.[8]

A letter from Sander to Roezl in the early 1880s, gives a flavor of Sander's life:

> My God, what a lot I've got to tell you and how shall I begin? Up to the 31st July I had to find £5,000 of foreign exchange and God knows how I did. The auctions went very badly . . . Yesterday there came in a huge consignment from the travellers in the Philippines, probably all dead — certainly a loss of about £600 to £800. Shortly before that three cases of Cypripedium from the same area arrived frozen. Fourteen days ago a ship carrying 177 cases of orchids went under. I damn near went mad. Work like a dog and not to know what's coming out of it. I hope against hope and fight. All my travellers except three are back home.
>
> You have no idea of my troubles. I have five greenhouses full already and goods outside waiting, and you coming in July.[9]

The cost and difficulty of obtaining many species, far from deterring the nurserymen and their collectors, was a large part of their attraction.[10] Sander had many aristocratic and wealthy clients, including both Baron Rothschild of Vienna and Nathaniel Rothschild of Aylesbury, as well as Lord Salisbury and the Dukes of Marlborough and Devonshire and Baron Sir John Henry Schröder, a wealthy merchant banker.[11] When the orchidmania reached its peak during the last quarter of the nineteenth century, men like these competed for orchids and would pay extravagant sums to be the first to own a new and rare species. The highest sum paid for a single orchid in late Victorian times was achieved at Protheroe and Morris' sale rooms in London (where many of Sander's plants were sold): Baron Schröder bid 610 guineas for a single plant of *Odontoglossum crispum Cooksoniae* (a guinea was £1.1s, i.e., 21 shillings), but a Mr. Peters "was determined not to be beaten" and paid the staggering sum of 650 guineas (roughly £298,000/$471,000 today) for a plant that "consisted of one old bulb and one fine new bulb with a leaf eight inches long."[12] For Sander, such sums were — of course — very welcome, but the publicity was even more valuable; newspapers reported the amounts paid when a new species or variety of orchid was rare (there was supposedly only one other plant of this variety of *O. crispum* in Britain when Peters made his bid), but soon afterwards further shipments would arrive and prices fell rapidly, at which point numerous customers of more modest means would rush to Sander's nursery to possess the new variety. Take, for example, the brilliant gold *Cypripedium Spicerianum*, an astonishingly beautiful orchid from India. The first one to be sold in London in the late 1870s fetched over £250 (about £105,000/$166,200 today); within a year the plants were selling for about two shillings each, a fall of almost 25,000% thanks to the massive increase in supply that resulted from competitive collecting.[13]

Given the potential profits, Sander found it worth hiring more collectors, of whom the German Wilhelm Micholitz (sometimes spelled Micholicz) was one of the most successful and who continued working for Sander until the outbreak of WWI.[14] Sander soon came to trust Micholitz, so in 1882 when another collector in the Philippines refused to divulge the exact location of a beautiful new *Phalaenopsis* (perhaps, Sander suspected, to sell them to a rival nursery), Micholitz was dispatched to find the flower. Micholitz told Sander he had braved a cholera epidemic in Manila, then left for the jungle where, being "up to my knees in mud" he had removed his boots, only to be "covered in leeches." He nevertheless managed to collect more *Phalaenopsis* than his rivals, adding that "when the area was exhausted . . . I went to the other side of the lagoon and there I found a new Phalaenopsis."[15]

Collecting from the wild was essential for Victorian orchid traders, partly to look for novelties to keep the buyers interested, but also because many of the prized tropical orchids were almost impossible to cultivate outside their native countries. William A. Stiles, editor of *Garden and Forest*, the first American specialist botanical and gardening magazine, noted that orchids were much harder to propagate than many other plants, which kept prices high. And even if you could pollinate them, it might be many years before they flowered (nineteen, in the case of *Laelia callistoglossa*; 21 for *Laelio-cattleya Veitchii*). Only a few cuttings could be taken each year and many species grew very slowly. As a result, the only way "to secure a garden of orchids" was

> to send collectors into tropical forests where masses of them have been forming for generations and bring them out alive. To carry them safely on the backs of coolies or mules, over mountain-trails and through malarious marshes, sometimes for weeks, until a stream is reached down which they can be floated to a seaport, is no light undertaking. And then the perils of the voyage, half-way round the globe perhaps, are still more serious. On deck the salt spray means death, and in the hold the close and heated air is often fatal, while every bruise is the beginning of active decay.[16]

Despite the orchid collectors' increasing understanding of where and how orchids grew, the wastage involved in collecting remained appalling; thousands of plants died in the warehouses before they could be shipped, thousands more on board, and often whole ships were lost.

Undaunted by such difficulties, Sander kept demanding more plants and complained endlessly of the cost of paying collectors, many of whom—especially Micholitz—felt unappreciated. He wrote to Sander from Singapore (May 5, 1889), commenting that "I received your letters of the 12th and 20th of March and I cannot say they particularly delighted me. It is always the same cant with little variation; how I am using too much money and not sending enough. Well, do you suppose I find novelties in the streets of Manila?" Micholitz's letters also reveal his opinions of the various "natives" he had to deal with; they were far from complimentary (and typical of prevailing European attitudes). He told Sander that Singapore was mainly inhabited by "Chinese and Malays, who are like the mosquitoes and suck the blood out of us whenever they can."[17] But he seems to have had a slightly higher opinion of the Papuans, commenting that although "they think no more about killing somebody as your cook does killing a fowl," they are handsome people, who "go about entirely naked and are the finest race of savages I have ever seen up to now."[18]

Lost Orchids

In the last decades of the nineteenth century, orchid-hunters scrambled to be the first to discover a new species, just as Europe's colonial powers were competing to carve up the remaining portions of the un-colonized world. Upon finding a rarity, collectors would gather as many specimens as they could, often leaving few if any behind, and when the new plant made its appearance in Europe's salerooms, the precise location from which it had come was—of course—kept secret.[19] Given the collectors' methods, it is hardly surprising that occasionally a new orchid would cause great excitement at auction, but when collectors were sent back to the tropics to find more, the species had vanished. The reappearance of "lost orchids" regularly created sensations in the orchid sale rooms and newspapers, but the London-based daily newspaper the *Standard* reported that the month of October 1891 would be "memorable in the annals of orchidology" because Protheroe and Morris had auctioned not one but two fabulous lost orchids in London.[20] These two remarkable sales, occurring so close to each

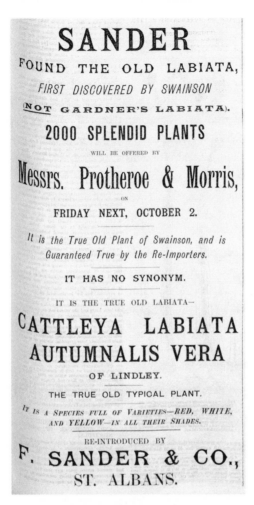

Figure 21. An advertisement from Sander & Co.
trumpeting the rediscovery of *Cattleya labiata*.
(From the *Gardeners' Chronicle*, 1891)

other, helped ensure that the romance and danger of orchid hunting would be
permanently lodged in the public's imagination.

The first sale, on October 2, was of the fabled *Cattleya labiata vera*, which
had been missing for over seventy years. In about 1818, a plant collector working in Brazil, William Swainson, sent some plants to the British botanist William
Hooker (father of Joseph, who would later be director of the Royal Botanic Gar-

dens, Kew), which included a new species of orchid. There's an oft-repeated but implausible story (which apparently originated with Joseph Paxton) that the orchid's tubers were simply shoved into a box of lichens as part of the packaging for the more interesting plants (unlikely because orchid tubers are generally too hard to create the cushion that such packaging is intended to provide).[21] Regardless of how it arrived, the new orchid proved not only to be a new species, but a new genus, christened *Cattleya* by John Lindley (who was becoming one of the world's foremost orchid specialists), to honor William Cattley in whose greenhouse it first flowered. The particular species that Swainson had sent, with a magnificent violet blossom, was named *Cattleya labiata* in honor of the prodigious and showy labellum. Lindley wrote that he was delighted to have a chance to compliment "a gentleman, whose ardour in the collection, and whose unrivalled success in the cultivation, of the difficult tribe of plants to which it belongs, have long since given him the strongest claim to such a distinction."[22]

The beautiful new orchid was much admired and collectors returned to the region near Rio de Janeiro where Swainson had supposedly found it, in the hopes of gathering more specimens. Collectors assumed it would be easy to find; as Frederick Boyle noted, they assumed the orchid "must be a common weed at Rio, since men used it to 'pack' boxes," but they were disappointed. There were plenty of similar *Cattleyas* there but they were not exactly the same; the original was hardier and easier to grow and it bloomed for longer. A further attraction was that Swainson's original species showed several attractive variations that could be hybridized with other species to produce seemingly endless new forms. Why not ask Swainson where they had come from, since he was alive and well, and living in New Zealand? Boyle reinforced the myth of *C. labiata*'s accidental discovery by suggesting that Swainson would have had no advice to give: "The orchids fell in his way by accident—possibly collected in distant parts by some poor fellow who died at Rio. Swainson picked them up, and used them to stow his lichens."

For many decades, cuttings from Swainson's original plant were the only source of this flower, and as one collector after another failed to rediscover it, it acquired a near-mythical status among orchid lovers, eventually becoming known as *Cattleya labiata vera*, or the "true" *Cattleya labiata*. According to Boyle, there was "not one" among the major orchid nurseries in Europe, "who has not spent money—a large sum, too—in searching for *Cattleya labiata vera*."[23] It would be seventy years before the orchid reappeared. All kinds of stories have circulated about how the "lost" orchid was found, my favorite being that it was spotted in the corsage of a lady at an embassy ball in Paris where an orchid enthusiast attached to the British legation saw it. He looked once, twice (in fact as often as etiquette would permit a gentleman to do). Convinced that this was the lost orchid, he somehow obtained the flower and an expert confirmed his guess as to its identity. The Holy Grail had at last turned up; the trail was followed back to the precise place where it had originated in Brazil and orchid fanciers were finally able to obtain fresh stocks.[24] A lovely story, but there doesn't appear to be any evidence for it. The prosaic truth (which has been in print since at least 1900) is that the dates of Swainson's original shipment had been misunderstood; the collection that included the original *Cattleya labiata* had not been sent from Rio but following an earlier trip to Pernambuco, in the northeast of the country (and about 1,400 miles from Rio). Once the original location had been identified, there were plenty of orchids to be collected. The *Orchid Review*'s editor, Robert Allen Rolfe, took Boyle to task for the inaccuracy of his "Lost orchid" story in 1900 and concluded that if the facts had been known, "the history of the 'lost orchid' would have been vastly different, and would probably never have been written at all."[25]

Rolfe's advice was ignored; the story of the "lost orchid" had already been told many times, and was to be repeated many more, not least because it helped generate publicity. The "true" *Cattleya* had been reintroduced with great fanfare by Sander in 1891 (who gave it the *vera* name). Large advertisements in popular magazines, such as the *Gardeners' Chronicle*, announced the reappearance of "the True Old Plant of Swainson, and is Guaranteed True by the Re-Importers," adding "IT HAS NO SYNONYM" (meaning the name was not shared by any other species) and "IT IS THE TRUE OLD LABIATA" and—just in case anyone were in doubt—the ad repeated that the 2,000 plants to be sold were nothing but "THE TRUE OLD TYPICAL PLANT."[26] Those loud claims were aimed at Sander's rival, Jean Jules Linden of Brussels (after whom the Ghost Orchid, *Dendrophylax lindenii*, was named), who claimed to have rediscovered the plant at about the same time.

However, one extraordinary orchid sale in a month was apparently not enough for Sander. Just three weeks later, the firm announced "the orchid sensation of the century," in the form of *Dendrobium phalaenopsis Schröderianum*, or the "Elephant Moth Dendrobe" as they named it, "the king of the genus," which the advertisement modestly proclaimed to be "the grandest and in every way the finest known Orchid." Nearly 1,000 plants of this species were to be sold on October 16 and the ad explained that "there are only seven plants in Europe, which are mostly in the princely Collection of Baron J.H.W. Schroder" and that these had all "come from two small plants originally introduced to the Royal Gardens at Kew."[27] And if rarity and beauty weren't enough to entice buyers, one of the rediscovered orchids was auctioned while growing on a human skull.

The orchid on the skull became a legend, but — as with so many other orchid stories — this one grew in the telling. At the time, the *Standard* reported that the species was not merely stunningly beautiful but "has the advantage of a romantic history, in evidence of which extracts from the letters and telegrams of the collector who despatched it were affixed to the catalogue." The reporter explained that "the native land of a new orchid is kept secret, as a rule, as long as possible," but this one came from New Guinea (just north of Australia). When the first consignment of plants made it to the small port from where they were to be sent home, the ship caught fire and every orchid was lost. The collector telegrammed Sander — "Ship burned, everything lost; what do?" [*sic*]. The brusque reply was "Return. Try again." The collector responded "Rainy season on," but according to the newspaper " 'Return — collect more,' replied the ruthless capitalists." Finally, the explorer returned and found the plant again at a new spot, where he recorded "what a glorious thing it is." However, as the newspaper reported, this beauty was discovered in a macabre setting:

> The spot was the burial ground of the savages — at least, the place where they deposited their dead — and the lovely flowers were rooted among skulls and human bones. The literal truth of this sensational story is beyond dispute, for one noble plant arrived growing out of a dead man's jaw. Its roots have overrun most of the face, and a fern of the Platycerum class clothes the base of the skull like a ruff. This extraordinary object was shown yesterday in a private room avoid a crush. The natives had a not unnatural reluctance to disturb the bones of their forefathers for the sake of weeds, but "when they saw the handkerchiefs, beads, looking glasses, and brass wire I offered, they did not trouble themselves any more about the souls of their ancestors."

The *Standard* added that the indigenous collectors had insisted that a golden-eyed idol be packed with the plants to protect them "and this divinity made the first 'lot' in yesterday's sale."

The reporter gave a few details of the plant's history, noting that it was not a new species, but "one of our rarest orchids," having been first collected on behalf of Kew by a Mr. Forbes in 1880, but that a Captain Broomfield in Australia had made drawings of a similar variety a little earlier, while "Baron Schröder also has a few plants," supposedly from the Southern Philippines. However, the New Guinea collection was both more beautiful and more abundant than any previous sample, so much so that "masses of *Cattleya labiata vera* — which had been the cynosure of all eyes in this same room but a fortnight ago — lay almost unnoticed." The top price paid for a *Dendrobium* that day was 28 guineas and by 4 PM, Sander and his auctioneers had realized £1,500 (about £639,00/$1 million today). And, despite being overshadowed by the newcomer, the *Cattleyas* fetched another £1,000, despite another £1,500 worth having been sold earlier in the month.[28] Clearly, a romantic tale or two did no harm to the sale prices.

The *Standard* did not give the name of the intrepid collector who had been ordered back by that "ruthless capitalist," Sander, but it was Micholitz and most of the details of how "the orchid sensation of the century" came to be found, clearly came from Micholitz's letters, hundreds of which survive at Kew and provide fascinating insights into the life of an orchid collector. When Sander ordered him back after the first shipment burnt, Micholitz responded that the nutmeg traders had left, the season being over, and "it would be very unsafe to go there when they have left":

> However should it be absolutely necessary to get them I would risk it, only I would have to take several well armed men with me and the plants would be rather dear at the same time [.] I must tell you I think it would be a matter of some difficulty to collect say 2–5000 of one kind, in the place itself there are only very few natives and they are at war with a neighbouring tribe.[29]

He closed by asking Sander to "write explicitly" with instructions, "also I beg you will not forget to renew my credit, you will bear in mind that I have lost absolutely everything." As we've seen, Sander's telegrams were admirably "explicit" and Micholitz returned. He wrote again, a few months later, describing the flower as "a glorious thing" and giving the details of exactly where it had been found:

> I saw the first on some rocks near the little village where I was staying, where they grew on the bare limestone between a great number of human skulls and bones — the natives do not bury their dead but put them in a kind of coffins [*sic*] and then place them on these solitary rocks [which] stand scattered about along the shore or beach and which can be reached or got at only at ebb tide.

This letter must have been circulated when Sander put the flower on sale, since it was obviously the source of the details in the *Standard*'s story, which were then repeated by many other newspapers. This letter went on, for example, to mention that indigenous people's reluctance to collect the flowers, "afraid the souls of the departed whose bones lay there bleaching in the sun, would resent it," and explained how Micholitz had been able to bribe them with "gorgeous handkerchiefs, looking glasses, etc," but he reassured Sander that "you need not be afraid[;] I shall send you no bones or skulls with them."[30] It is surprising, therefore, that a skull did in fact appear in the sale room, but it clearly did the publicity campaign no harm; two years later, Joseph Godseff, who was Sander's manager, suggested to Micholitz that he should return and collect more skulls with orchids growing on them, since they were sure to generate considerable interest. Micholitz was not amused and replied that he had "no particular desire to leave my own skull behind me, to figure in the collection of some Papuan . . . I value it far too much to risk it for the sake of affording merriment in the sale-rooms."[31]

The orchid on the skull was, naturally, an irresistible one for a journalist with a passion for orchids like Frederick Boyle, and he first wrote about the rediscovery of the Elephant Moth Dendrobe in an article that described a typical sale, enlivening his description by contrasting it with anecdotes of more famous sales, including the one on October 16.[32] Boyle must, however, have realized that he'd rather wasted the *Dendrobium* story by mentioning it so briefly, so he returned to the tale a few years later in a piece called "The Story of an Orchid" (a title he recycled for several other stories about other species) in which, rather characteristically, he started to embroider and embellish the already remarkable tale. For example, he says that species was first "sent to Kew by Fortune in 1857," but the celebrated plant collector Robert Fortune never went to Australia or

New Guinea; the orchids he collected, like most of his other plants, came from China. Boyle says this single plant remained a "special trophy" at Kew for many years, where it was successfully multiplied. Sir Joseph Hooker himself gave small pieces to the well-known orchid collector John Day, to Schröder, and to the nurserymen Veitch and sons, who sold theirs to Schröder. When Day's collection was dispersed after his death, Schröder bought his specimen too, thus ending up with all three known specimens of the variety that had been named after him. Numerous collectors sought the plant in vain until Sander decided "to wait no longer upon chance. He studied the route of Fortune's travels, consulted the authorities at Kew, and, with their aid, came to a conclusion," and dispatched his collector (but of course had he actually consulted *Fortune's* records, Micholitz would never have got near the plant). Boyle added mysteriously that

> it is universally understood that Micholitz discovered the object of his quest in New Guinea. If that error encouraged the exploration of a most interesting island, as I hear, it has done a public service. And the explorers have not wasted their time. They did not fall in with Dendrobium Schroderianum, because it was not there.

So where did the flower come from? Boyle would only say that "very shortly now the true habitat will be declared. Meantime I must only say that it is one of the wildest of those many 'Summer Isles of Eden' which stud the Australasian Sea."

Who could blame Boyle for keeping the location secret given that, as he wrote, "the man of sense who finds a treasure does not proclaim the spot till he has filled his pockets, nor even, if it may be, till he has cleared out the hoard," and Sander was doubtless entitled to a return on the discovery that had begun with his cunning reconstruction of "Fortune's" travels.[33]

However, these cunning efforts to cover the collector's tracks were entirely unnecessary; it was a matter of published record that *Dendrobium phalaenopsis Schröderianum* had been collected by a Scottish naturalist, Henry Ogg Forbes (the *Standard* had got Forbes' name right in its original report), in 1882, in Timor Laut, a small group of islands between Australia and New Guinea. After he returned, Forbes published an account of his travels in the *Proceedings of the Royal Geographical Society* (1884), which clearly recorded where he had been. His (rather modest) botanical haul was presented to London's Linnean Society in the same year, and his report mentioned he had "found a handsome and, I imagine, a new species of orchid," details of which were published the following year in the widely read *Botanical Magazine*, with a beautiful illustration and a botanical description by Hooker, who wrote, "The specimen here figured is one of the few botanical prizes secured to England by Mr. Forbes during his adventurous expedition to Timor-laut."[34]

There was no mystery about the plant's collector, location, or date, so why all the (pardon the pun) skullduggery? Why did Forbes become Fortune (who died in 1880, before the orchid was ever seen in Britain)? Perhaps the magazine's publisher made the mistake (they did, after all, manage to misspell Boyle's first name), although that's unlikely since "Fortune" occurs more than once in the article. Perhaps Boyle had simply written "Fortune" by mistake (a Freudian slip, perhaps, given the orchid's value? — and he did correct the name when the piece was republished in 1901), but the spurious secrecy over the plant's location seems like a deliberate attempt to add an aura of mystery. In addition, Boyle's date is totally wrong (Forbes was six years old in 1857, which would have made him a very precocious botanical explorer indeed), but pushing the plant's first discovery back helped make this another "lost" orchid, allowing Boyle to embellish his tale with the fantasy of a twenty-year search. The other curious aspect to this story is that the lost orchid was not even a species; Hooker dryly commented that he could "detect no difference between the . . . Timor-laut [*sic*] plants" collected by Micholitz and the original *Dendrobium phalaenopsis* collected in Australia, which had been named and published a couple of years earlier. As far as Hooker was concerned, Micholitz's prized specimens didn't even rank as a subspecies (which is what the orchid's third name, *Schröderianum*, indicated), he simply lumped the Timor Laut plant with the Australian one, referring to them all by the original name, *Dendrobium phalaenopsis*. So, in Hooker's view, the "orchid sensation of the century," far from being a "special trophy," was merely the confirmation of a new locality for an existing species.

The various embellishments to Micholitz's story (which hardly seemed to need embellishing) suggest the hand not only of Boyle but of Godseff, who often wrote to collectors asking for publicity material in the form of hair-raising tales of adventure. He usually got fairly "dusty answers," particularly from Micholitz. In later years, Sander's son Fritz (as Frederick junior was known) came to regard Godseff as a "rogue" who cost the family firm thousands, but couldn't persuade his father to dismiss him.[35] However, Boyle not only didn't share Fritz's suspicions, he described Godseff as "my guide, comforter and friend, Joseph Godseff" (in his collection of essays *About Orchids: A Chat*), and in *The Culture of Greenhouse Orchids* (1902), Boyle asked his reader rhetorically, "some may ask what are my credentials for offering advice upon the culture of orchids?" but assured them that "Mr. Joseph Godseff has examined every page of the work, and he allows me to affix his imprimatur."[36]

Cannibal Tales

Commercial considerations were, of course, one major reason for embellishing the various tales of "lost orchids"; hair-raising adventures and frequent loss of life prompted—and justified—high prices. But as those prices always fell as soon as further supplies arrived in Europe, fresh novelties were needed, preferably accompanied by skulls, idols, tales of cannibals, and other tropical myths. In the article where he first told the skull story, Boyle commented casually that "only last week we heard that Mr. White, of Winchmore Hill, has perished in the search for *Dendrobium ph. Schroederianum*."[37] (Given that G. T. White was exhibiting orchids at the Royal Horticultural Society's show in London in the same month as he was supposedly being devoured in New Guinea, it is possible that Boyle may have exaggerated reports of his death.) Despite the lack of evidence for cannibalism, newspapers kept repeating the stories. London's *Daily Mail* newspaper told its readers that the Solomon Islands (in the Pacific) were home to a stunning species of multicolored orchids, but "there cannibalism is still all but unchecked. Orchid hunters who have ventured there aver that the natives when they offer human sacrifices to their gods load the victims with garlands of these gorgeous blossoms," although the hapless collector might take some posthumous comfort from the alleged cannibals' habit of decking their victims with orchids, with the flowers' "colours growing richer and deeper hued with his spurting blood."[38]

The American monthly, *Scribner's Magazine*, published a long illustrated piece on orchids that included many similar tales, of which "the latest tragic story" was of a collector forced into marriage with an African chief's sister in order to secure a prized orchid from Madagascar, *Eulophiella elisabethae*.

The collector who was "compelled to face the dread alternative of death or matrimony" was a Monsieur L. Hamelin, who claimed that he had befriended a

Figure 22. Eulophiella elisabethae, the orchid that Monsieur Hamelin claimed to
have married an African chief's daughter to obtain. (From *Lindenia,* 1891)

local chief, Moyambassa, becoming his blood-brother, thus making it possible
for him to explore the jungle accompanied by a royal guard that included the
king's brother-in-law. Unfortunately,

> in these dense woods the dreaded *Protocryptoferox Madagascariensis* lies in
> wait for his prey upon the branches of the tree where the Eulophiella grows,

crouching among the great tufts of the orchid with its tall, arching spikes of white and lurid-purple flowers. These carnivores are no respecters of rank, and one of them sprang on and rent to death, the husband of the princess.

Hamelin then learned that as he had caused the man's death, he must "appease the spirit of his relative by being greased and burned"; the alternative was to "assume his position and responsibilities in the royal household. The collector promptly decided to brave the love of the princess and become brother-in-law of the chief."[39]

According to *Scribner's* "the confiding editor of one of the great horticultural journals of Europe, 'sees no reason to doubt the veraciousness' of the victim's narrative." That phrase identifies the journal in question as London's *Gardeners' Chronicle*, which had published a letter from Hamelin (from which all the details in the *Scribner's* story were drawn), prefaced by the comment that "astonishing as the statement is, there is no reason to doubt its veraciousness"). The *Eulophiella* story really took off when it was reported in the *Standard* under the title "M. Hamelin's adventures in Madagascar," which consisted of a letter to Sander & Co., describing the orchid's discovery. The *Standard* corrected the scientific name of the native "lion" (a species of civet cat known as a fossa) to *Protocrypta ferox* (it was mangled by Hamelin and the mistake was repeated in *Scribner's*). Despite this correction, the story aroused skepticism, not least because Hamelin's motivation for telling the tale was fairly obvious: he wrote that "an amateur paying one hundred shillings for a plant would not cover the cost" of the hardships he had endured. And, in case potential buyers were tempted to wait until fresh importations brought the cost down, he made it clear that the remaining plants were

> under the special care of my brother in blood, Moyambassa, until such time as I may want them . . . and at least several years must elapse before these small plants will be large enough to gather. Amateurs may trust that no plant of this species can or will be imported."

Nor should anyone think of repeating his journey; he recalled "the sad fate of three inexperienced and unfortunate collectors, who, not knowing the customs of the country, died — two from fever, the last by the sword of the natives."[40]

The *Gardeners' Chronicle* version included similar comments that made it

From a Sketch by W ELLIS   P. 380.

Audience at the Palace Antananarivo.

Figure 23. The royal palace in Antananarivo where Hamelin claimed his
"blood brother" Moyambassa was guarding the remaining orchids.
(From William Ellis, *Three Visits to Madagascar*, 1858)

absolutely clear that nobody else would be able to get the orchid: "My brother-in-law's will is absolute in the country of the Eulophiella, and from three years' exploration I do not believe this plant exists anywhere else in this terrible country." And Hamelin claimed that a "terrific storm at sea" had destroyed many of the orchids before he could get them home, further inflating their value (and making it even less likely that other supplies would be found).[41] It is surely no coincidence that the same issue of the Gardeners' Chronicle included a large advertisement from Sander, promoting a forthcoming auction of Eulophiella elisabethae, which stated: "To avoid useless correspondence, M. Hamelin desires us to state that every plant saved of his entire importations are in our hands, and no plants can be acquired elsewhere, or from any other source whatever, and this we can fully guarantee."[42] Sander's company was a regular advertiser in the paper, which probably made it even harder for their editor to publicly cast doubt on Hamelin's "veraciousness."

The Standard's editors, perhaps being less dependent on advertising from orchid dealers, expressed a note of caution when they published their account; there was a short piece in the same issue that commented on how interesting Hamelin's story is, but added that "with a reserve which does credit to his commercial instinct, the French traveller does not indicate too precisely the scene of his adventures." The Standard also observed that Madagascar's dominant tribe were fairly well subdued by the French and that the island's nominal ruler was Queen Ranavalona III, whose "husband [is] too good a Christian to indulge in any of the peculiarly disagreeable practices to which the French orchid collector alludes so complaisantly." Nevertheless, the Standard's notes of caution were ignored by the many newspapers that picked up the tale. The Liverpool Echo repeated the story, as did the Nottinghamshire Guardian, which quoted Hamelin's claim that "no other European can travel there and escape death," adding that one of the London orchid houses had subsequently sent a collector there, but "he has not been heard of since."[43] And the Yorkshire Herald told its readers how Hamelin had not only "won a dusky bride," but secured the remaining orchids "from all poaching on the part of brother depredators — or, more euphemistically, plant-hunters."[44] The story traveled as far as Australia, where the Melbourne Argus reprinted Hamelin's letter from the Standard (but again without the cautionary commentary) and in the United States, Garden and Forest (whose editor, William A. Stiles, wrote the later and longer piece in Scribner's), credited Hamelin with discovering E. elisabethae, of which he "had previously sent only

Figure 24. Phalaenopsis stuartiana, one of the illustrations that accompanied
W. A. Stiles' account of Hamelin's adventures. (From *Scribner's Monthly*, 1894)

three plants to Europe." The plants that reached London had apparently been
sold by Sanders for 3 to 5 guineas each and "M. Hamelin assures us that he collected every plant that he could find worth collecting."[45]

None of the journalists who repeated Hamelin's tale seems to have bothered
to read the *Standard* over the few weeks after the original tale appeared. Had
they done so, they would have read a letter from a Mr. R. Baron, of Antananarivo, Madagascar, whose twenty years' residence there made him doubtful about
Hamelin's claims, "certain it is that many—nay, most, of the statements made by
this gentleman are somewhat imaginary."[46] (Baron also wrote to the *Gardeners'*
Chronicle, which published his correction, accompanied by the rather improb-

able claim that they had never believed Hamelin in the first place.[47]) A couple of days later, the *Standard* published a further letter from Mr. P. Weathers, the English representative of Linden's company in Brussels—*L'Horticulture International Société Anonyme*—one of Sander's European rivals. Weathers pointed out that not only had Hamelin not discovered the plant (it had been described and published the previous year), but he hadn't even collected the plants he was trying to sell. Lucien Linden, son of the firm's founder, had sent to Madagascar a collector called Sallerin, who died there, after which Hamelin somehow acquired his plants. Hamelin brought them to Marseilles and tried to sell them, but when Linden discovered that several hundred plants had already been sent to Sander in England, the Belgian orchid grower felt that "the good faith which is usually considered to exist between collector and employer had been broken" and refused to buy them. Weathers concluded that

> it will, therefore, be seen that M. Hamelin has not only told curious stories of his adventures in search of the plant, but that he has absolutely no right to claim the credit of its discovery—this latter point being the one which interests us mostly, as introducers of new plants of over fifty years' standing.[48]

Lucien Linden himself had already published a detailed account of the full story in June 1893 (before either the *Standard*'s or *Gardeners' Chronicle*'s stories had appeared), describing Hamelin as a "jay who has adorned himself with the plumes of the peacock." Hamelin had apparently contacted Linden and Son in March 1891 to inform them of Sallerin's death and offer his services as a collector. By that time Linden had not only received *Eulophiella elisabethae* but had seen it flower; excited by Sallerin's discovery, they sent Hamelin a drawing so that he would know what to look for (Hamelin was in the shipping business and doesn't appear to have had any previous plant collecting experience). Linden claimed that "On the 5th. October 1892, M. Hamelin wrote to us that he had recognized the plant on receipt of our pamphlet, and intended to send us examples," but when he finally contacted them it was to claim "AS THIS SPECIES WAS ONE OF MY DISCOVERIES, *will you state what you will offer for the entire stock, for I have completely destroyed the plant in its native habitat.*" (The claim that orchid collectors regularly destroyed rarities to retain exclusive control and keep prices up is often repeated in histories of orchidmania, but this is the only nineteenth-century reference to the practice I have come across.) Linden placed all this on record, for the sake of "our unfortunate collector Sallerin, who is no longer able to defend himself, that which belongs to him. *It will be admitted as right and proper that the false and erroneous statements which have been made should be corrected.*"[49]

Perhaps the various journalists who spread Hamelin's tale didn't read Baron's and Weathers' letters, nor Linden's more detailed correction, or maybe they decided not to let the facts get in the way of a good story. Whichever it was, Hamelin's farfetched tale rapidly became one of the standard orchid-hunting stories. The *Singapore Free Press* recycled it a few years later, under the head-line "Orchid Hunting: A Dangerous Pastime," telling its readers that Hamelin "had to go through the ceremony of being made 'brother-in-blood' to King Moyambassa in order to penetrate the interior of this island, an honour which nearly cost him his life." The article (apparently based on details from an issue of *Harmsworth's Magazine*) explains that wherever orchid-collectors gather one would hear "remarkable stories of hair-breadth escapes and of appalling suffering, but unhappily in too many cases the daring orchid hunter never returns to tell the story." The writer retold the story of *Dendrobium Schröderianum* (without mentioning the species or its collector by name), which had been found "rooted among the bones and covering the ghastly remains with a mantle of gorgeous flowers." As a result, "Many of the plants could not be torn off the bones, and one skull was brought home to England with an orchid firmly rooted in the brain cavity and growing out of the jaw." And, predictably, the tale of *Cattleya labiata*'s finding, disappearance, and reappearance got another outing too:

> An orchid, of quite new and unknown species, arrived in the packing in which some foreign plants were sent home. No one knew where it came from, and for a long while it continued unique. Orchid hunters sought everywhere for it, but not until seventy years later was it found.[50]

The coincidence of the true *C. labiata* and *D. Schröderianum* being auctioned within the same month in London surely helped to link their—much embroidered—stories. Lost orchids became a recurring thread in the rich fabric of the high-Victorian orchid tale, along with breathtaking prices being offered by competing aristocratic patrons. Rivalries between collectors and importers became entangled with images of flowers on skulls and fanciful tales of "dusky brides" in distant jungles. Even the great European myth of cannibalism, which had been used since the times of Columbus to justify the imperial enterprise, got bound up with orchids. (To his credit, Boyle acknowledged that every tribe in New Guinea accused its neighbors of cannibalism but that, to the best of his knowledge, they were all lying.[51]) The numerous ways in which orchid facts became orchid fictions are all-too-easy to see, but it's worth asking why: what did the orchid stories mean to those who told and listened to them?

Orchids came to embody the romance and opportunity of empire, of the outlandish, almost fanciful places the readers might one day aspire to visit, at least in their imaginations. Orchid-hunting stories allowed the listeners to take imaginative possession of their own tropical paradise. One late-Victorian orchid catalogue struck a curious balance between the romance and practicality of orchids when it assured potential growers that

> orchids as a rule, are not more costly than other select plants; the culture
> of them is nearly as simple as that of hyacinths and other high-class bulbs;
> and there is no reason why every man who has a conservatory, and who will
> lay out a little money judiciously, and treat his plants kindly and lovingly,
> may not render it, with these orchid-treasures, as gay as that famous Saracen
> Alhambra.[52]

The evocation of Spain's Alhambra, Granada's fabulous Moorish palace, reveals that orchids remained exotic even after their Victorian democratization. Despite the increasing ease of growing them, they were part of a persistent orientalist fantasy that encouraged Europeans to daydream of everything from humid, leech-infested jungles to enthralling dark-skinned temptresses while they watered their greenhouse of imported orchids. To own an orchid was, in some senses, to annex a little piece of distant foreign lands, to hold a trophy of empire in your own garden.[53]

However, the high-water mark of orchidmania, the end of the nineteenth century, also marked the end of imperial expansion; there were almost no more territories left unclaimed, no unexplored lands to conquer, perhaps no more new orchids to find? The cannibal tales are an aspect of a shift that was taking place in orchidmania during the final decades of the nineteenth century. In the 1890s, *Scribner's* tried to re-recreate a sense of adventure around orchids by reminding its readers that "nearly all the orchids we see have been torn from their homes, and carried away captive," a fact that "helps to thicken the atmosphere of mystery which surrounds them; and this mystery is not mitigated by the thrilling narrative so often heard of hardships endured by the men who go in quest of them."[54] Yet even as the magazine offered its readers the latest orchid romances, they had to acknowledge that "the orchid-hunter's life is not as hazardous or romantic as it once was" and to admit that "a well-told adventure may be an effective advertisement for a new plant." It was gradually becoming possible to grow the plants successfully, with the result that "the present demand is not for a few rare plants, but for greater quantities of the more popular kinds, not only for the decoration of private green-houses, but to produce cut flowers in the market."[55]

The same sense that the great days of orchid-hunting were over is apparent in Boyle's account of a typical orchid sale of the same period. He admitted that "excitement does not often run so high as in the times, which most of those present can recall, when orchids common now were treasured by millionaires." There is a wistful sense in much of the journalism of this period that an era is ending; Boyle observed that "steam, and the commercial enterprise it fosters, have so multiplied our stocks, that shillings—or pence, often enough—represent the guineas of twenty years back."[56] The steamships had brought the jungles closer, so that orchids became just another imperial commodity, no more exotic than tea, which had once been expensive and mysterious, a carefully guarded Chinese secret, but now came routinely from British controlled plantations in India. The romance of trade and exploration aboard sailing ships was being lost as even the fast clippers were increasingly replaced by regular steamships, and telegraphs and railways contributed to an annual shrinking of the globe. The exaggerations and poetic flourishes that characterize late-Victorian orchid-hunting tales suggest a desire to re-enchant orchids, to bring back the mystery and danger of earlier days, not just to the orchids themselves, but also to the humid, exotic, and sensual world they symbolized. As the century drew to a close, it would increasingly be fictional orchids, rather than supposedly true tales of collecting, that best allowed this re-enchantment, infusing orchids with new and ever-stranger meanings.

7

Savage Orchids

In 1894, the British magazine the *Pall Mall Budget* published "The Flowering of the Strange Orchid," a curious "tale of an orchid enthusiast" called Winter-Wedderburn, a small, drab suburban bachelor in his fifties whose only excitement in life was "one ambitious little hothouse" full of orchids. He complains to his housekeeper (a distant female cousin who is never named in the story) that nothing ever happens to him; he yearns for adventure of some kind. As he finishes his tea and toast, in preparation for catching the train to yet another London auction, he admits to envying Batten, the man who collected the orchid tubers he plans to bid on, even though the hapless collector is now dead. Still, thinks Wedderburn, what a life he led:

> That orchid-collector was only thirty-six—twenty years younger than myself—when he died. And he had been married twice, and divorced once; he had had malarial fever four times, and once he broke his thigh. He killed a Malay once, and once he was wounded by a poisoned dart. And in the end he was killed by jungle-leeches. It must have all been very troublesome, but then it must have been very interesting, you know—except, perhaps, the leeches.

· 129 ·

Two wives, four bouts of malaria, and a dead Malay! Even death by leeches seems to Wedderburn a small price to pay for such a rich and varied life.

As the story's narrator tells us, there was "a certain speculative flavour" in buying orchids. Despite the advances in cultivation and shipping, most orchids reached Europe as dormant tubers or stems, a "brown shrivelled lump of tissue," giving the purchaser little idea whether the plant would even grow, much less what it will prove to be. (The real-life orchid expert Benjamin Williams made the same point: "There is a risk in buying imported plants, since there are many that do not turn out as represented."[1])

When Wedderburn returns from his sale he is "in a state of mild excitement" (despite not having acquired a wife, leeches, or malaria), "he had made a purchase," an unknown bulb that might prove to be something new—and entirely unexpected. As Wedderburn tells his housekeeper:

"That one"—he pointed to a shrivelled rhizome—"was not identified. It may be a Palaeonophis [sic]—or: it may not. It may be a new species, or even a new genus. And it was the last that poor Batten ever collected."

His housekeeper is distinctly underwhelmed by the tuber, pointing out that it is ugly and looks "like a spider shamming dead." But Wedderburn is caught up in the aura of mystery that surrounds the rhizome:

"They found poor Batten lying dead, or dying, in a mangrove swamp—I forget which," he began again presently, "with one of these very orchids crushed up under his body. . . . Every drop of blood, they say, was taken out of him by the jungle leeches. It may be that very plant that cost him his life to obtain."

Wedderburn plants it and nurtures it and finally, the strange orchid flowers. Before he even glimpses the bloom, he notices a "a rich, intensely sweet scent, that overpowered every other in that crowded, steaming little greenhouse" and then saw the blooms. "He stopped before them in an ecstasy of admiration":

The flowers were white, with streaks of golden orange upon the petals; the heavy labellum was coiled into an intricate projection, and a wonderful bluish purple mingled there with the gold. He could see at once that the genus was altogether a new one.

Figure 25. "The last that poor Batten ever collected"—Mr. Winter-Wedderburn
examines his new purchase. (From H. G. Wells, "The Flowering of
the Strange Orchid," *Pearson's Magazine*, 1905)

The heat and smell are almost overpowering and as he tries to open a window he faints. When he doesn't appear for tea, his housekeeper goes looking for him:

> He was lying, face upward, at the foot of the strange orchid. The tentacle-like aerial rootlets no longer swayed freely in the air, but were crowded together, a tangle of grey ropes, and stretched tight, with their ends closely applied to his chin and neck and hands.
>
> She did not understand. Then she saw from one of the exultant tentacles upon his cheek there trickled a little thread of blood.

In an instant she realizes that it was not jungle leeches that sucked Batten's blood from his veins, but the orchid itself. With great presence of mind, she

Figure 26. "She saw from one of the exultant tentacles upon his cheek there trickled a little thread of blood." (From H. G. Wells, "The Flowering of the Strange Orchid," *Pearson's Magazine,* 1905)

smashes a window before she too is overcome and manages to drag Wedderburn out of the orchid's clutches and into the fresh air.

"The Flowering of the Strange Orchid" was, as you may have guessed, written by H. G. Wells, who was rapidly becoming a well-known short-story writer (even though real fame would come only after the publication of *The Time Machine* in 1895).[2] "Strange Orchid" is clearly rooted in the tales of supposed real-life orchid collecting described in the previous chapter, but being freed from the need to even pretend to be telling the truth allowed Wells to create an even richer evocation of the lure and danger of orchids. However, fictional orchids are not a digression from the history of real ones; on the contrary, it is only by understanding how the tales of real and fictional orchids grew alongside each other that we can fully understand the roles orchids have played in our culture; science and fiction were intertwined well before the term "science fiction" was coined.

Although it is one of the best stories of its kind, "Strange Orchid" is far from unique. Killer orchids were among many lethal plants that stalked the suburban greenhouses and the imaginations of their cultivators in the late nineteenth century. By the 1880s, dangerous orchids and other flowers seemed to lurk everywhere, ready to devour unsuspecting humans at every opportunity. This was clearly a popular (and successful) genre, which survived well into the twentieth century, but there seem to be no stories like these before the late 1860s. The fin-de-siècle literary orchids took the imperial romance and eroticized it in new ways, often linking orchids to homosexuality and other kinds of potentially dangerous sexual encounters. Why did orchids become such sexy, decadent killers in the late nineteenth century? Answering that question helps us understand how orchids would reshape our visions of nature in the twentieth.

Long Purples and a Forked Radish

Like so many other flowers, orchids made their debut in Western literature in the works of William Shakespeare. In Act IV of *Hamlet* (1623), we learn that poor mad Ophelia has drowned herself in the weeping brook, and

There with fantastic garlands did she come,
Of crow-flowers, nettles, daisies, and long purples
That liberal shepherds give a grosser name,
But our cold maids do dead men's fingers call them:

By "long purples" Shakespeare meant orchids, probably a species of Orchis. The "grosser" name used by shepherds would have been one of those we learned in earlier chapters, such as dog's, goat's, or fool's stones (i.e., testicles) and such plants were, as we have seen, often assumed to have aphrodisiac properties and were associated with satyrs and other lustful creatures. There is a painful, typically Shakespearean irony in the image of the virgin Ophelia adorning herself in these tokens of desire, as she despairs of Hamlet's love.[3] The other common name for this orchis, "dead men's fingers," suggests one of the "handed" species, such as *Orchis serapias*, whose roots look like fingers rather than like testicles; Shakespeare was clearly more interested in filling Ophelia's final scene with images of sex and death (amorous shepherds and "cold maids," aphrodisiac "stones" and dead men's fingers), than he was in getting his orchid species straight. Shakespeare's imagery would echo through many later literary evocations of orchids, as the endless retelling of the spurious eighteenth-century myth of Orchis demonstrates, and this entangling of sex and death was undoubtedly one source of killer orchid stories like Wells'.

Orchids are, of course, not the only flowers Shakespeare invested with symbolic meaning. In one of *Hamlet*'s earlier scenes, Ophelia babbles increasingly incoherently about the symbolic meanings and virtues of plants, describing rosemary as "for remembrance" and pansies for thoughts of love, and enigmatically instructing her brother to wear the herb of grace, or rue (*Ruta graveolens*), "with a difference." For most of history, humans have projected all sorts of meanings onto plants, but while flowers might signify purity or passion, they weren't traditionally imagined as exhibiting these virtues or vices themselves. Indeed, when a poet wanted an image of dull passivity, something incapable of taking action or feeling emotion, "vegetable" was often the adjective reached for. Consider another line of Shakespeare's (*Henry IV, Part II*, Act III, scene II), when Falstaff disparages the old, lying Justice Shallow:

> "I do remember him at Clement's inn, like a man made after supper of a cheese-paring: when he was naked, he was, for all the world, like a forked radish, with a head fantastically carved upon it with a knife."

The force of the insult (that nothing appears to hang between the two forks of Shallow's legs) is intensified by the comparison to a radish; a man may honorably be many things, but he should never be a vegetable.

However, even if flowers (or radishes) had been imagined as lacking passions in Shakespeare's day, one might assume that by the end of the eighteenth
century, Linnaeus' sexual system of plant classification would have transformed
their image. The sexual imagery in the Swede's writings were so vivid that some
of his contemporaries were outraged. William Smellie (key compiler of the first
Encyclopaedia Britannica) complained about the "alluring seductions" of the Linnaean system and he accused Linnaeus of taking his analogies "beyond all decent
limits," even claiming that the Swedish naturalist's books were enough to make
the most "obscene romance writer" blush.[4] Yet, if we look more closely at Linnaeus' analogies, it often seems that—despite having brought the carnal possibilities of the plant world so energetically to life—the relationship between
plants and human sexuality was very different for men and women. The shock
experienced by people like Smellie was largely aroused by thoughts of the effect
such imagery might have on women, who were imagined to be as pure and unsul-
lied as the flowers they were so naturally suited to arranging, painting, or embroi-
dering. A woman could indulge in a little light botany, ideally as a way of teaching
her children to appreciate the beauties of nature and the wisdom of its creator,
but neither she nor her innocent offspring should be reading about Linnaeus'
"marriages of plants," which were really botanical orgies when phalanxes of floral
husbands "embraced" numerous brides in their richly colored and scented bou-
doirs of petals, with no concern for Christian norms of monogamous matrimony.

Although Linnaeus saw the plant world as richly erotic, the flowers were
never imbued with *masculine* desires, intentions, or agency. Linnaeus privately
categorized his wife using his own plant classification system, calling her "my
Monandrian Lily," that is, as a flower traditionally associated with virginity, possessing only one husband.[5] While men should not be flowers, fruits, or vegetables, it seems that women could be, precisely because of their imagined lack
of animal lusts. Linnaeus was neither the first nor last to associate flowers and
women; a century after his death, a British orchid-growing manual would comment that "plants are marvellously docile. . . . and with orchids especially, as with
women and chameleons, their life is the reflection of what is around them."[6] Numerous writers ascribed moral purity—and passivity—to women while denying
them fully human desires (or rights). As a result, the force of Linnaeus' metaphor always seems to be that while plants are sometimes like people, people are
not really like plants, apart from women (who, it seems, are therefore not really
people!).

The supposed similarity between women and flowers lies at the heart of one of the earliest killer plant stories, Nathaniel Hawthorne's story "Rappaccini's Daughter" (1844), the tale of Giovanni, a medical student at the University of Padua "very long ago," who gets lodgings overlooking the garden of a mysterious doctor called Rappaccini, who specializes in growing poisonous plants.[7] The student becomes entranced by the doctor's beautiful, flower-like daughter, Beatrice, who is described as "beautiful as the day, and with a bloom so deep and vivid that one shade more would have been too much." When Giovanni first sees her he thinks of her "as if here were another flower, the human sister of those vegetable ones," and when the doctor asks Beatrice to take sole charge of the most spectacularly sinister plant in the garden (because it's too toxic for him to approach even with mask and gloves), his daughter addresses the flower as "my sister."[8] As Giovanni falls in love with the girl, he discovers that she has become so toxic through regular contact with the flowers that nobody can touch her and live. Giovanni procures a powerful antidote but when Beatrice takes it, she dies. It's an intriguing, rather Gothic story, that blends the girl and the flower into a single, sinister image of untouchable toxicity, but it is striking that the girl is depicted as having no more agency or willpower than the flowers do. She may be (very literally) a *femme fatale*, but she's no seductress, merely the passive recipient of her father's obsession and of the young man's fascination.

By contrast, in the later stories—like Wells' "Strange Orchid"—the plant is actively malevolent. Wedderburn's housekeeper compares it to "a spider *shamming* dead" and when Wedderburn shows her the newly emerging aerial rootlets, she compares them to "fingers trying to get at you." As these rootlets grew longer, they "reminded her of tentacles reaching out after something; and they got into her dreams, growing after her with incredible rapidity." There is no mistaking the orchid's active pursuit of its human prey, nor its cunning. These qualities are precisely what makes supposedly innocent flower so sinister; it has become imbued not merely with the qualities of an animal, but with those of a predatory animal. However, Wells' orchid—distinguished by its malevolent intent (it has "exultant tentacles") as much as by its flowers—is far from unique. Another example was "The Purple Terror," by Fred M. White, which appeared in 1896 in the *Strand Magazine* (which first published the Sherlock Holmes stories). Most of the killer plant tales appeared in popular magazines like the avidly read products of the nineteenth-century industrialization of publishing (discussed in chapter 4). White's story featured as its hero the implausibly named Lieutenant Will

Figure 27. Beware of women wearing orchids: Zara attracting the
Americans' attention with "the wreath of flowers around her shoulders."
(From Fred White, "The Purple Terror," *Strand Magazine*, 1896)
Credit: Author's collection.

Scarlett, of the US Navy, described as "rather a good specimen of West Point
dandyism," who is sent on an expedition to deliver an important dispatch while
avoiding anti-American rebels.[9]

Scarlett and his companions stop in a small village on the first night of their
journey and while drinking in a wine-shop, Scarlett's eye is caught by Zara, a
stunning Cuban dancing girl, but more particularly by "the wreath of flowers
around her shoulders." Zara's boyfriend ("a bearded ruffian") is jealous of Scarlett's attention to the girl, but being an enthusiastic botanist, the American is
most interested in her garland:

> The flowers were orchids, and orchids of a kind unknown to collectors any-
> where. On this point Scarlett felt certain. . . . The blooms were immensely
> large, far larger than any flower of the kind known to Europe or America, of
> a deep pure purple, with a blood red centre.

That blood red center is the first hint that there may be something sinister
about these orchids, but not the last:

As Scarlett gazed upon them he noticed a certain cruel expression on the flower. Most orchids have a kind of face of their own; the purple blooms had a positive expression of ferocity and cunning. They exhumed [*sic*], too, a queer sickly fragrance. Scarlett had smelt something like it before, after the Battle of Manila. The perfume was the perfume of a corpse.

If only Scarlett had read Wells' "Strange Orchid" (as White clearly had), that "sickly" scent might have alerted him, but this hint of danger is no deterrent to a serious orchid-lover. When the dancing girl's clearly jealous boyfriend, Tito, offers to show Scarlett where the orchids grow, the American is too preoccupied by orchidmania to question the Cuban's motives; he even ignores the fact that Tito slaps Zara across the face to stop her from trying to warn him of some danger connected to the flowers. "All Scarlett's scientific enthusiasm was aroused. It is not given to every man to present a new orchid to the horticultural world. And this one would dwarf the finest plant hitherto discovered."

Off they go and when the orchids are sighted, their green tendrils look "like a huge spider's web, and in the centre of it was not a fly, but a human skeleton! The arms and legs were stretched apart as if the victim had been crucified." Many would turn back, but not our hero! He accepts Tito's dubious explanation that the skeleton must be that of a hapless plant collector who got trapped in the vines "like a swimmer gets entangled in the weeds." And when Tito proposes they camp beneath the orchid vines in a clearing that is strewn with bones (including another human skeleton), Scarlett still doesn't smell a rat. Even Tito's decision to sleep outside the circle of bones doesn't alert our clueless hero.

Scarlett is awoken by his favorite mastiff howling in pain and finds the dog's dead body, bloody and lacerated, and "scattered all about the body was a score or more of the great purple orchid flowers." Then he sees the orchid vines, trailing down to the ground; their "broad, sucker-like termination was evidently soaking up moisture." As in Wells' story, these are vampiric orchids, whose rapidly coiling tendrils (a weird variation on the aerial rootlets of epiphytes) seek blood. The Americans are almost killed but swift reactions and sharp knives save the day. After threatening to leave him tied up in the middle of the clearing while they complete their mission, Tito confesses all and they take him away for trial (presumably on a charge of attempted murder by orchid), and probable execution.

Stories like these seem connected to the perception that orchids were the "highest" (most evolved) plants, which may explain why their characteristics

Figure 28. Sharp knives were sometimes needed to
stop the orchid hunters from becoming the orchid's prey.
(From Fred White, "The Purple Terror," *Strand Magazine*, 1896)
Credit: Author's collection.

were sometimes combined with those of carnivorous plants to create an imaginary intermediate between animals and plants. That idea was certainly present in the strange tale "Kasper Craig" (1892), by Maud Howe (daughter of the prominent abolitionist Julia Ward Howe, who wrote "The Battle Hymn of the Republic"). The tale concerned a penniless young man, Leonard Ebury, who meets Kasper Craig, a wealthy orchid collector, at a flower show. Craig has eccentric theories about affinities and harmonies between different living things, with higher people being naturally attuned to higher plants (a theory that echoes Bateman's view of the natural social hierarchies that should order plants and people; see chapter 4). He illustrates these for Ebury by applying them to the women they see at the flower show. Ebury is offered unexplained work and is uncertain until he sees a beautiful young girl, Mary Heather, who works for Craig. Attracted by the girl, he accepts the job and later finds Mary at work on a painting of an extraordinary orchid, which Craig explains is his latest attempt to re-create the evolutionary missing link between animals and plant, an orchid that has animal-like characteristics. The plant is already carnivorous and sensitive, and it has definite periods of sleep and wakefulness (the sleep of plants being another topic Darwin explored[10]). Ebury's task is to travel to the jungles of South America to collect another species of rare orchid that will become part of Craig's experiment, but he becomes convinced that the orchid Mary is painting is sucking the life out of her; she gets paler and paler as the orchid acquires an increasingly rosy color. Ebury destroys the orchid, an action that mysteriously liberates Mary from Craig's sinister influence.

Howe's "Kasper Craig," White's "Purple Terror," and Wells' "Strange Orchid" clearly have themes in common. The blood-draining orchids are reminiscent of vampires (although Bram Stoker's classic *Dracula* was not published until 1897, after these three stories had appeared) and feasting on blood is hardly proper behavior for a plant. In each case, the tropics where the orchids come from are imagined as sinister places, where white manliness seems endangered by uncanny flowers, savage animals, and alien, dark-skinned people. Wells' story evoked all this while hinting at a fear that the subjugated tropics might re-invade their colonizer's homes via the garden and greenhouse. (What might be called colonizer's guilt was a theme Wells would later explore in *The War of the Worlds*, 1898, where humanity's helplessness before the technologically superior Martians is explicitly compared to that of Tasmania's Aborigines, "swept out of existence in a war of extermination waged by European immigrants."[11]) These stories, like much of the literature of the late nineteenth century, are tinged with melancholy, a sense that the sun may indeed be setting on the great European empires because there are no blank spaces left on the map, no new lands to be annexed.[12] Other fin-de-siècle tales have a similar mood, but the orchid stories gave a distinctively biological twist to the wider mood; like Wells' orchid, White's has animal-like qualities (the tendrils are compared to the arms of an octopus, and move with a speed no real plant could match), and the plants seem conscious, able to perceive their surroundings and to stalk their prey. The uncanny quality of these stories is partly created by a blurring of the boundaries between plant and animal (exactly Craig's goal), as if the order of nature itself were starting to unravel. How and why did man-eating and apparently sentient plants become a recurring theme, apparently for the first time in Western literature, during the final decades of the nineteenth century? Surprisingly, these lurid tales were closely connected to the scientific study of flowers.

Queer Flowers

In 1884, the *Popular Science Monthly* carried a fascinating article called "Queer Flowers," which commented on the cruelty of nature. When a cliff collapses and kills someone, we can put it down to mere accident, but

> when a plant is so constructed, with minute cunning and deceptive imitativeness, that it continually and of malice prepense [aforethought] lures on the living insect . . . to a lingering death in its unconscious arms, there seems to be a sort of fiendish impersonal cruelty about its action which sadly militates against all our pretty platitudes about the beauty and perfection of living beings.[13]

This plant is not some monster from Wells' imagination, but a sundew, a small plant common in bogs and marshes. It has tiny hair-like spikes on its leaves, surmounted by glistening tips that look like drops of water and catch the light — hence its name. The article's writer was Grant Allen, a popular novelist and journalist who was deeply interested in science; he knew Darwin personally and was one of the most successful Victorian popularizers of his work (and wrote what seems to be the first biography of Darwin).[14] Wells would later acknowledge Allen's influence on him, praising the "aggressive Darwinism" that pervaded his

FIG. 1.*
(*Drosera rotundifolia.*)
Leaf viewed from above ; enlarged four times.

Figure 29. The apparently murderous sundew,
one of the most popular of Darwin's plants.
(From Charles Darwin, *Insectivorous Plants*, 1875).

popular natural history and acknowledging that Allen "had a very pronounced streak of speculative originality." In his autobiography, Wells drew some parallels between his life and ideas and those of Allen (who was his near neighbor in later years).[15]

One of many distinctive aspects of Allen's writing was that it was about plants. Then (as now), most of those who wrote about Darwinism focused on animals; the idea that nature was red in tooth and claw was easily grasped when animals were actively attacking and devouring each other. It was harder to illustrate natural selection at work in the meadows and hedgerows, but Allen took up the challenge, and the insect-eating carnivorous plants like the seemingly malicious sundew provided a perfect example. Darwin had pioneered the study of such plants, writing a book on them (*Insectivorous Plants*) in 1875. Following on from the work on orchids, *Insectivorous Plants* was the final part of his three-pronged assault on the public's understanding of plants. In between these two, Darwin explored his third theme, that plants were far more mobile than most people realized. His argument began with an article in a scholarly journal "On the movements and habits of climbing plants" (1865), which became a book a decade later. Once seen through Darwin's eyes, plants proved to be far more like animals than most people had realized. They not only moved but, as he showed in *Climbing Plants*, they appeared to sense where there was a support they could scramble up, not only detecting but reacting to their immediate surroundings. As we have seen, *Orchids* showed—amongst other things—how the complex adaptations ("contrivances") of plants allowed them to manipulate insects to ensure cross-pollination. And, of course, Darwin had also used them to show how there could be contrivance without a contriver, rejecting all thought of divine intervention and demonstrating instead that the benefits of avoiding self-fertilization were enough to explain the breath-taking variety of the orchids' forms. *Insectivorous Plants* caught the public's imagination even more vividly, as it dramatized how the vegetables sometimes turned on the animals, devouring instead of being devoured.[16] Yet despite Darwin's fame, not many of his contemporaries had the time or patience to read even the *Origin*, much less the drier and more technical botanical works; it was popular writers and journalists like Allen who appropriated these vegetable carnivores and transformed them into the dramatic centerpieces of widely read works. It was important that the plants they wrote about be exciting ones, and described in attention-grabbing prose; as Allen admitted to Darwin, he wrote for money, "I earn my whole livelihood by writing for the daily or weekly press. . . . I can only give to science the little leisure which remains to me after the business of bread-winning for my family is finished."[17]

The sundew was among the favorites of Allen, who noted that Darwin wrote "a learned book about it" and the poet Algernon Swinburne had "addressed an ode to it," but "not because it is beautiful, or good, or modest, or retiring, but simply and solely because it is atrociously and deliberately wicked." Those lustrous droplets attract insects who, when they land to enjoy what look like drops of nectar, find themselves caught in a very sticky situation. Gradually, the sundew envelops the trapped insect, folding its tentacle-like leaves around its victim like a sea anemone. Allen used vivid adjectives to describe the plant: assigning it "murderous propensities" embodied in the "cruel crawling leaf" that secreted a digestive juice over the insect and "dissolves him alive piecemeal in its hun-

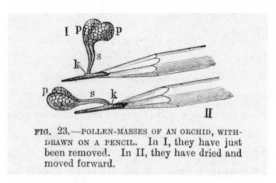

FIG. 23.—POLLEN-MASSES OF AN ORCHID, WITH-
DRAWN ON A PENCIL. In I, they have just
been removed. In II, they have dried and
moved forward.

Figure 30. Grant Allen illustrated Darwin's experiment of using
a pencil in place of an insect's proboscis to show how the
orchid's pollinia moved as their "glue" dried.
(From *The Story of Plants*, 1895)

dred clutching suckers." He admitted that this "mute tragedy" moved him to the point that he sometimes tried to "release the fresh flies from their ghastly living but inanimate prison," but there were too many plants for him to visit them all.[18] Yet the sundew's malice was easily explained: the bogs and marshes where they grow have poor soils, with little nitrogen in them; the plant has adapted to this environment by ingesting the nitrogen in insect bodies. Around the world, numerous other plants—including the various pitcher plants (species of the genera *Sarracenia* and *Nepenthes*), bladderworts (species of *Utricularia*), and the Venus fly-trap (*Dionaea muscipula*)—also have adaptations that allow them to ingest insects and adapt to similar conditions. Allen used them all to dramatize Darwin's ideas and illustrate them in action.

Orchids also figured regularly in Allen's writing alongside the sundews and carnivorous plants, even though the orchids don't share their insect-devouring habits. Allen used orchids to describe the complexity of pollination and the mutual dependence of plant on insect that had created their fantastic array of shapes, "apparently out of pure wantonness, but really in order to ensure fertilization by the oddest and most improbable methods."[19] Books like Allen's must have linked orchids and insectivorous plants in the public's mind, and his vivid descriptions may have inspired Wells, who had himself been taught biology by Darwin's most accomplished (and belligerent) publicist, Thomas Henry Huxley. It's certainly easy to see the germ of "Strange Orchid" in Allen's writing when he described an insect struggling helplessly in the arms of the sundew: "it is impossible not to feel a little thrill of horror at this battle between the sentient and the insentient, where the insentient always wins—this combination of seeming cunning and apparent hunger for blood on the part of a rooted, inanimate plant."[20]

Given that orchids are not carnivorous (only one species—unknown in Allen's time—is suspected of carnivory[21]), it might seem strange that orchids became the subject of several killer plant stories. Admittedly, killer orchids as subjects are relatively rare; most of the killers are based on real insectivorous plants and take the form of giant Venus fly-traps, massive pitcher plants, or monstrous sundews.[22] However, Howe, Wells, and White were not the only writers to plant a few deadly orchids amid their more obviously dangerous cousins. Short stories (some of which will be discussed later) include Marjorie Pickthall, "The Black Orchid" (1910); John Blunt, "The Orchid Horror" (1911); Edna Underwood, "An Orchid of Asia" (1920); James Hanson, "Orchid Death" (1921); Oscar Cook, "Si Urag of the Tail" (1923); Bassett Morgan, "The Devils of Po Sung" (1927); Gordon England, "White Orchids" (1927); John Collier, "Green Thoughts" (1931); and Wyatt Blassingame, "Passion Flower" (1936); and Marvin Dana built an entire novel around lethal orchids in *The Woman of Orchids* (1901).[23] And the protagonists of other killer plant stories are often orchid hunting when they meet their fate, as in Clark Ashton Smith's creepy "Seed from the Sepulchre" (1937), which concerns an alien plant that can germinate only within a living human's skull (an image that strongly recalls the story of the *Dendrobium phalaenopsis* on a skull that was sold in London; see previous chapter).[24]

Why should the poor, innocent orchid have so much imaginary blood on its petals? There's a clue in Grant Allen's writing when he described orchids as being among "the most advanced families" of plants, and therefore "cleverer and shiftier than all others."[25] The idea that plants might possess the hitherto unsuspected (and very animal-like) property of being *cunning* was derived from Darwin, but became much more vivid when writers like Allen reinterpreted his botany for new audiences. Just look at the dates of the killer plant stories (above)—it is surely no coincidence that not one predates Darwin's evolutionary reinvention of plants, beginning with *Orchids* (1862), but they really take off once Allen and others began their imaginative appropriation. The first germs of the killer orchids are there in Darwin's descriptions of their real-life counterparts, such as the Brazilian bat orchid, *Coryanthes speciosa* (chapter 4), in which the flower's labellum has been enlarged into a pitcher-like bucket (like a carnivorous pitcher plant) into which the hapless bee falls and pollinates the flower as it crawls out—just one of many examples of orchids outwitting their pollinators. Similarly, when Darwin described *Cypripedium* (the genus known as Lady's Slipper orchids), he observed that a small insect could crawl in but not out and so "the labellum thus acts like one of those conical traps with the edges turned inwards, which are sold to catch beetles and cockroaches in the London kitchens."[26] Some Australasian orchids, such as *Pterostylis trullifolia* (the Trowel Leaved Greenhood) and *P. longifolia*, actually do imprison their pollinating insects; the labellum springs up when they land on it and, as Darwin wrote, the insect "is thus temporarily imprisoned within the otherwise almost completely closed flower. The labellum remains shut from half an hour to one hour and a half, and on reopening is again sensitive to a touch."[27] Again, the insect has to brush against the pollinia before it can escape. Such devious orchids could be readily be imaginatively transformed into killers.

However, Darwin's influence on the killer plant stories goes well beyond a few suggestive descriptions of particular orchids' behavior. He had also been responsible for a popular fascination with insectivorous plants.[28] When his book on the topic was reviewed by the highly scientific and serious journal *Nature*, the writer commented: "Even the newspapers have discussed the anti-vegetarian habits of some vegetables in the light, airy, and philistine manner in which they are wont to approach 'mere scientific' subjects."[29]

Newspaper reviews, magazine articles, and popular books didn't just spread Darwin's theories, they reimagined them, ensuring that his name linked orchids and insectivorous plants in the imagination of a public who increasing saw the natural world in terms of deadly competition and struggle, rather than as a divinely ordained harmony; the killer plant stories were one aspect of a much broader shift in humanity's view of nature. However, there was also a subtler aspect to his influence; in his work on climbing plants he stressed how in climbers like the passionflower the plant's "tendrils place themselves in the proper position for action," as if waiting for an opportunity to climb. Such plants demonstrate, he argued, "how high in the scale of organisation a plant may rise," and he compared their "admirable" tendrils to the arms of an octopus (the same simile used by White in "The Purple Terror").[30] In a letter to his son, Darwin went as far as to describe tendrils as "more sensitive to a touch than your finger; & wonderfully crafty & sagacious."[31] The idea that plants were *sagacious*, possessed of acute perception or judgement, was fundamental to the new Dar-

winian view of plants. Darwin was effectively claiming that plants possessed intentions, that they had agency. This was so startling that it took a long time to
for humans—including scientists—to accept. Popularized plants, in both fiction
and nonfiction, played an important role in allowing people to imagine for the
first time that plants might really possess plans, might wait for opportunities, so
that they could exploit insects and each other. After Darwin, plants possessed
"crafty" strategies for survival, reproduction, and everything else—and they also
possessed the means to achieve them—but without popular writers, including
those who wrote fiction, most of us might never have known how much plants
had changed.

Darwin's plant books were part of his wider strategy to break down the barriers that seemed to exist between the plant and animal kingdoms. Since evolution implied that all living creatures must at some point have had a common
ancestor, that hypothetical organism must have been neither animal nor plant,
but must have eventually given rise to both. It was therefore important for him
to demonstrate that there were no insurmountable barriers between the kingdoms; plants must possess—in however slight a measure—the same abilities as
animals. And, since nobody had ever seen a new species arise, it was vital to Darwin's persuasive strategy to find real examples that could easily be imagined as
small steps on the road to complex organisms. Plants like the Venus fly-trap were
able to detect the difference between a fly, or a small piece of meat, and a grain of
sand or small pebble; this ability to discriminate could be used to illustrate one
very small but plausible step on the long, slow ascent that took a single-celled
creature to full human consciousness.[32]

Plants changed dramatically once they were viewed through evolutionary
eyes; they were more exciting (and potentially more popular) subjects to write
about, and Grant Allen was not the only writer to seize on the remarkable new
qualities that Darwin had discovered in them. Another science journalist, John
Ellor Taylor, curator of the Ipswich Museum and founder of various local scientific societies, wrote a popular Darwinian botany book with the fabulous title *The
Sagacity and Morality of Plants: A Sketch of the Life and Conduct of the Vegetable
Kingdom* (1884). Had it been published forty years earlier, such a title would undoubtedly have identified the book as a satire (at best), but evolutionary botany
had now made the lives "and conduct" of plants a serious scientific subject. As
Taylor wrote, botany "no longer consists in merely collecting as many kinds of
plants as possible, whose dried and shrivelled remains are too often only the

caricatures of their once living beauty"; instead, "It is now a science of Living Things, and not of mechanical automata, and I have endeavoured to give my readers a glance at the laws of their lives."[33]

Concerned that some of his readers might object to his title, by arguing that only conscious beings can be said to possess morality, Taylor quoted Darwin; the tip of a radicle (the plant's first root) "acts like the brain of one of the lower animals; the brain being seated within the anterior end of the body, receiving impressions from the sense-organs, and directing the several movements."[34] As a result of Darwin's insights, a new language has been developed to describe the "novel relationships" between plants and other organisms and Taylor asserted that "whether we believe in the consciousness of plant-life or not, this language almost implies such a belief." There was, Taylor acknowledged, an inevitable consequence of accepting evolution: every feature of the higher animals, including human intelligence, must have been built along "an unbroken scale of minute steps from the 'minutest animalcule' up to Shakespeare."

Establishing that plants possess the earliest rudiments of intelligence was not, Taylor thought, too difficult; after all, they have their likes and dislikes just as we do—some orchids thrive in heat, others need the cool. He moved onto more contentious ground when he argued that

> hosts of common plants constantly perform actions which, if they were done by human beings, would at once be brought within the category of right and wrong. There is hardly a virtue or a vice which has not its counterpart in the actions of the vegetable kingdom. As regards conduct, in this respect, there is small difference between the lower animals and plants.

While Taylor did not believe that plants have responsibility for their actions, he was convinced there was an unbroken continuity from the lowest plants to the highest animals.

When it came to describing the complex ways in which plants manipulate their pollinators (which Taylor described as "Floral Diplomacy"), the orchids took pride of place as the most developed and specialized of flowers. Taylor argued that "orchids all over the world are of the most ingenious character" and acknowledged Darwin for bringing their ingenuity so vividly to the world's attention that "his Various Contrivances by which Orchids are fertilised by Insects, reads like a romance."

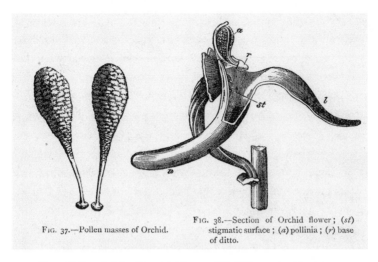

FIG. 37.—Pollen masses of Orchid.

FIG. 38.—Section of Orchid flower; (*st*) stigmatic surface; (*a*) pollinia; (*r*) base of ditto.

Figure 31. Popularizing Darwin's ideas could include reusing his pictures while surrounding them with more accessible words.
(From J. E. Taylor, *The Sagacity and Morality of Plants*, 1884)
Credit: Author's collection.

Just as Darwin had done, Taylor encouraged his readers to conduct their own, hands-on research, urging them to carefully thrust "the conical point of a black-lead pencil down the throat of an Orchid flower." As we have seen, being able to join in was a key reason for the popularity of Darwin's botany and Taylor claimed that in the few years since Darwin, the scientific understanding of plants had grown more than it had in the whole century before him (an exaggeration perhaps, but a forgivable one). This explosion in botanical knowledge had been possible because "the humblest observer" could study such things as the mechanisms for ensuring cross-pollination, and as a result would "render loyal and genuine service to the cause of science."

Readers who chose to join Taylor in repeating Darwin's experiments would, he averred, increase their "admiration for these aristocrats of the floral world!" The rather paradoxical message that working with these "aristocrats" was now open to all was perhaps a reflection of late nineteenth-century Britain becoming more socially mobile than it had ever been, but at the same time becoming more acutely aware of social distinctions (precisely because social mobility prompted uncertainty as to which class people properly belonged). Most of Allen's, Darwin's, and Taylor's readers would have been part of that rapidly growing but ill-defined group, the middle-class: a group that was happy to look down on working people but increasingly reluctant to look up to the aristocracy, whose loose morals and laziness became a regular focus for middle-class satire and invective. (Look, for example, at the ways aristocrats were portrayed in Victorian novels; Anthony Trollope offers numerous examples of feckless, dim-witted sons of Lords with nothing to offer but a title and an impoverished estate.) Taylor regularly pursued analogies between plants and people and imposed a characteristically middle-class morality on the vegetable kingdom. He noted that flowers and fruit consume the energy that leaves generate, which is why plants may go many years without flowering and why abundant fruiting seasons may often be followed by dearth. Plants became members of the Victorian middle-class, who had read Thomas Malthus' celebrated *Essay on Population* and therefore knew it was their duty to defer marriage until they have saved enough: "The desire 'to found a family' is as manifest among plants as among men!" Taylor noted. "Otherwise what means this slow accumulation of energy on their part?"[35]

In the work of popular, middle-class writers like Allen and Taylor, Britain's wider social shifts were reflected in the garden, where Bateman's floral ancient regime (chapter 4) had given way to a more fluid sense of which people and plants might be well suited. As a result, even the most showy floral aristocrats, the orchids, could be reimagined as potential members of the blossoming middle class (which is, of course, rather self-contradictory, but being a member of the Victorian middle class required, above all, the ability to believe as many as six impossible things before breakfast).

In Taylor's tenth chapter (tellingly entitled "Social and Political economy"), he described the tubers of flowers, such as "those of the various Orchids," as a strikingly thrifty "store of starch, laid by for next year, saved out of last summer's vegetable earnings." And the epiphytic orchids were also complimented on the soundness of their political economy, when Taylor observed that they rely on the plant they grow upon, meaning they have few leaves and can save their floral energies for their blossoms. (Taylor misunderstood the relationship between the epiphyte and its support, which is not a parasitic one, but it was a common misunderstanding at the time and made his moral clearer.) "Remember," Taylor told his readers, "how lavishly the flowers of Orchids are called upon to expend"—like middle-class gardeners whose care with every penny ensured they would have enough money to fill a gorgeous greenhouse.[36]

People had been poisoning each other with plants—in fiction and reality—for many centuries, but what makes late nineteenth-century writing (fiction and nonfiction) distinctive is that the plants seemed to be developing murderous intentions of their own. Whether or not one really believed that plants were "moral" or "sagacious," by the late nineteenth century it was becoming increasingly hard to think of them as entirely distinct from animals; Darwin and his reinterpreters had produced a subtle but pervasive blurring of the lines between plants and people, creating a vague unease that was expressed in killer plant stories. As Taylor observed, people had long been used to thinking of plants in anthropocentric terms—as Linnaeus did, but:

> If modern botany had done nothing besides abolishing these crude views,
> it would have a claim to gratitude; but it has done more—it has taught us to
> regard plants as fellow creatures, regulated by the same laws of life as those
> affecting human beings themselves!

This captures the impact of Darwinism very crisply; it was now possible not merely to see ourselves in plants, but to see the plant in us.[37] We might even learn from the orchids, perhaps to be more thrifty or more cunning.

Creation and Consolation

Allen and Taylor were not the only ones to describe plants as purposeful. M. C. Cooke's *Freaks and Marvels of Plant Life: or, Curiosities of Vegetation* (1881), was another example of a writer using the carnivorous plants to try to interest readers in "a somewhat unpopular subject." However, *Freaks and Marvels* was published by the Society for Promoting Christian Knowledge, so Cooke had to deal circumspectly with the morality of the sundew and its creator. He cited experiments (including one by Francis Darwin, Charles' son) that proved that the insects that the carnivorous plants consumed were not killed gratuitously, but really did nourish the plant. So, despite the apparent cruelty, the effect was benevolent; fewer flies, and well-fed plants. And this was what a Christian ought to expect, after all:

> Wanton destruction, or wasted energy, are not the probabilities which would
> suggest themselves to the mind of any one who has devoted himself to the

study of the phenomena of life, nor would they elevate our conception of the All-wise Creator.[38]

What Cooke seems to have found slightly trickier was to fully acknowledge his intellectual debt to Darwin, while placating his orthodox audience. Fortunately for him, it was commonly accepted in Victorian times that scientific facts and theories were two different things, largely independent of one another.[39] Cooke was therefore able to remind his readers that Darwin was an "accurate and indefatigable observer" and that even the controversial naturalist's "bitterest foe has never accused him of distorting, or misrepresenting facts, for the benefit of any theory whatever."[40] Cooke's approach was far from unique; one of the reviewers of Darwin's orchid book had denied it was in any way "controversial, or even theoretical: facts, and facts only, form its basis."[41] This emphasis on the accuracy of Darwin's observations and experiments allowed conventional Christian gardeners to read his work on insectivorous plants, climbers, and orchids as collections of fascinating facts, while largely ignoring the theory of evolution.

As Darwin's readership grew, one of the more surprising results of his softening of the boundaries between plants, animals, and people — between humans and nature — was that some of his readers found consolations in evolution, particularly those who found they could not fully share Cooke's belief that the study of plants revealed them to have been created by the benevolent God of Christianity. Grant Allen made the striking observation that if one believed the "murderous" sundew to have been specially created by God, it would be hard to believe in its creator's benevolence; by contrast "it is quite a relief that we are able nowadays to shelve off the responsibility upon a dead materialistic law like natural selection or survival of the fittest."[42] The appeal of what might be called the consolations of evolution is perhaps the most unexpected of Darwin's impacts on his fellow Victorians.[43] The long-term effect of his orchid-based "flank movement" on traditional Christianity is evident in the way Allen wrote about the birds, bees, and orchids. For example, in several of his essays Allen refers to the natural object he is examining as the "text" on which he will "preach." Like the famous evolutionary popularizer, Thomas Huxley, Allen's botanical essays mimicked the popular form of a sermon to draw lessons from a common plant or flower.[44] However, Allen's message was not one to be heard from any Victorian pulpit:

The old school of thinking imagined that beauty was given to flowers and insects for the sake of man alone: it would not, perhaps, be too much to say that, if the new school be right, the beauty is not in the flowers and insects themselves at all, but is read into them by the fancy of the human race.

For Allen, the "whole loveliness of flowers" ultimately depended on "all kinds of accidental causes — causes, that is to say, into which the deliberate design of the production of beautiful effects did not enter."[45]

For some Victorians, the impersonal rigor of natural selection might be relentless, even brutal, but it was easier to understand than trying to imagine how an omnipotent, benevolent God could take a close, personal interest in each of our lives, yet seemed willing to allow cruelty to pervade creation. Why did the insect suffer in the sundew's "ghastly living but inanimate prison"? Why did so many Victorian children die so young? Why was there evil in the world? For some people at least, it was "quite a relief" to think that while God might have made the laws of nature, their impact on individual lives was a matter of pure chance.

After Darwin, orchids — like other flowers — were reimagined in many ways, but seldom as straightforward evidence of God's compassionate design. An interesting illustration of this shift can be found in a curious little pamphlet called *Feeble Faith: A Story of Orchids* (1882). Its author (identified only as "T. F. H.") clearly knew his or her orchids and began the book with a fairly detailed account of collecting in the tropics, where paid collectors paid "troops of natives to wander through the wildest districts" to hunt for orchids. Their trophies are brought back to London, often perishing en route, where they are auctioned:

> The interest of the purchaser is twofold: not only that of trying to make these hopeless-looking bits of vegetation grow again under his careful treatment, but also the hope of finding among his treasures, when after one or two years' of cultivation he has the good fortune to flower them, some variety possibly not known before, or if known, of rare occurrence and of extraordinary value.

This is clearly the world of Wells' Mr. Wedderburn, but the author offers a true story of how, while beginning a "small collection," he or she had met a Mr. M, "a buyer employed in a merchant's office in the city," who might have been a

FEEBLE FAITH,

A STORY OF ORCHIDS.

BY

T. F. H.

London:

HODDER AND STOUGHTON,

27, PATERNOSTER ROW.

Price Fourpence per dozen, or 2s. per hundred.

Phalænopsis Amabilis.

Figure 32. After Darwin, collecting orchids could be seen as a distraction from religious matters, whereas nature study had previously been widely regarded as a pious investigation of God's handiwork. (From T. F. H., *Feeble Faith, A Story of Orchids*, 1882) Credit: Cambridge University Library.

real-life Wedderburn, with his dull job, suburban garden, and three small green-houses full of orchids.

The author and Mr. M became friends, but lost contact because—as the author discovered—Mr. M had become seriously ill. She sent him a letter of sympathy and a religious tract ("Getting Saved," by a Methodist minister, the Reverend Mark Guy Pearse, 1874), in the hope that Mr. M's soul might be saved, should his illness prove serious. The result was that Mr. M saw the light. For years, his love of orchids had distracted him from God, with Sundays increasingly spent in the greenhouse instead of church, but the Reverend Pearse's pamphlet showed him the error of his ways. During his illness, his family had brought some orchids to his bedside to cheer him up,

> but when he caught sight of them, he quickly put out his hand and motioned for them to be removed, exclaiming, "No, take them away! they are now nothing but dross!"[46]

Saved from the temptations of orchids, he died peacefully a few days later. The poor flowers, once the perfect illustration of God's benevolent design, had become the epitome of worldly distraction from spiritual matters. In this one instance, at least, Darwin had outflanked the natural theologians more comprehensively than he might ever have imagined.

8

Sexy Orchids

Orchids are not just savage, they are sexy, too; sex and death, life's beginning and end, are coiled together in a single plant. Whatever the reasons for inventing the pseudo-classical myth of Orchis, its survival surely reflects the effectiveness with which it captured the orchid's paradoxical associations: divine beauty mingled with lowly lust, uncontrolled sexual passion leading to the brutal removal of the source of the lust. The orchid's role in our cultural imagination for centuries was so disturbing for the English art critic John Ruskin that he attacked what he described as the botanists' preoccupation with "obscene processes and prurient apparitions," and rejected the very idea that flowers were sexual. Many Latin and Greek names (notably the orchid's) were, he argued, too vulgar to be used ("founded on some unclean or debasing association") and so he suggested that orchids be renamed the Ophryds (*Ophrydae*) from the Greek word for an eyebrow, "on account of their resemblance to an animal frowning."[1]

Ruskin's suggestion was not adopted, of course, but anyone who was disturbed by the sexual origins of the orchid name might have taken some comfort from the fact that at least the mythical Orchis was a man; for centuries, European culture imagined lust as something only men experienced (recall Linnaeus' image of his wife as a virginal monogamous lily) and so women remained largely untainted by the "obscene processes and prurient apparitions" of the botanists. Yet one of the striking things about the dangerous orchids that emerged in the late nineteenth century is that they were not just cunning, cruel, and sexy, they were also conspicuously female — seductive *femmes fatales*, whose emergence reflected Victorian anxieties over shifts in both sexual morality and women's place in society.

In White's "Purple Terror," for example, the whole adventure begins with the beautiful Cuban girl, Zara, whose seductive dancing turns her wreath of orchids into "a trembling zone of purple flame" (interestingly the word "seductive" made its debut in English in the late eighteenth century — in a poem addressed to a flower).[2] White's imaginary orchids themselves are sinister, with their expression of ferocity and their "queer sickly fragrance," but their near-fatal allure is interwoven with that of the woman herself. Sexually predatory women became staple figures in cheap, mass-produced fiction around the turn of the twentieth century, which featured an endless series of "bad girls," often identified by their overuse of perfume. In many orchid stories, an almost overpowering, erotic scent created a miasma that dissolved the boundaries between women and orchids (all at around the same time that seductive women were first referred to as "man-eaters," or "vamps," i.e., vampires).[3] In Wells' "Strange Orchid," Wedderburn notices the flower's overpowering "rich, intensely sweet scent" before he even sees the blossom; like a seductive woman deploying her perfume to attract the male's attention, the orchid almost lures the innocent bachelor to his doom.

As the nineteenth century gave way to the twentieth, the intensely perfumed orchid often metamorphosed into an orchid-woman, particularly in popular fiction. For example, take John Blunt's story "The Orchid Horror" (which first appeared in a 10¢ pulp magazine called *The Argosy*, September 1911). The story is rooted in the now-familiar genre of nineteenth-century orchid-hunting stories, but takes off in unexpected directions when its hero, a man called Loring, is approached by a strange, almost corpse-like man, whose "yellow skin was drawn tight over his cheek-bones; there was far too little flesh on his limbs to make him an attractive figure." The sinister stranger invites Loring to see a wonderful collection of exotic plants in a nearby house, and when he enters its conservatory, he sees "Orchids, nothing but orchids, tens of thousands of the flowing, multi-shaded flowers, hung from the walls and ceiling of the place."[4] However, the real attraction emerges from among the flowers:

Figure 33. The "Malignant Flower" exhibiting its mastery over the great white hunters. (From *Amazing Stories*, September 1927) Credit: Collection of George Morgan.

> Before my eyes the rows of orchids bent apart. Slowly, with an exquisite grace;
> it was like the waving open of a lane in a wheat field, caused by the wind. Yet
> there was no wind . . . Wider became the opening of that aisle in the blossoms'
> close ranks. Suddenly at its end I saw — the collector's daughter.

The flowers sway in the absence of any breeze, as if under the command of the
woman who is never named in the story; Loring refers to her simply as his "Goddess of the Orchids" and immediately falls in love with this ravishingly beautiful woman, whose affinity with the orchids is so strong that she moves "with the
languid grace of the swaying flowers." She claims to return his love, but sadly
cannot leave her elderly father unless she can first bring him the only orchid
species missing from his collection, *Cattleya Trixsemptia*. (This is an imaginary
species, but the name is interesting: the genus *Cattleya* is, as we have seen, a real
one associated with fabulous "lost" orchids; the imaginary species name seems
to be built from the Latin suffix *-trix*, that signifies the feminine, in such words
as aviatrix, or executrix — with perhaps a hint of "trickery" as well; and "*Semptia*"
is probably derived from the Latin "*semper*," always. So the orchid's name means
something like "the eternal tricky female.")

Loring, who has never so much as picked a daisy before, sets off to South
America on his fairy-tale quest. He meets a professional orchid hunter en route
who is after the same flower and knows where to find an entire field of them.
However, when he learns that Loring's quest has been inspired by love, he asks
a curious question, whether he was "introduced to the lady of your choice by
a rack-boned skeleton of a man with yellow, deep-burning eyes — sort of an
upheaval-from-the-crypt sort of chap?" When Loring confirms this, the orchid
hunter reveals that the Goddess of the Orchids is well known among his profession; she is the one who's obsessed with orchids — she has no father — and
he refers to her as "a plant fiend" (has she a fiendish obsession with plants, or is
she a fiendish plant herself?). Lust has lured numerous collectors to their deaths
in search of the fabulous *Cattleya*; her cadaverous sidekick is the only one who
returned, orchidless but bound to her, desperate to win her love (or perhaps to
release himself from her spell?) by finding someone who can actually bring her
the mysterious orchid. Loring refuses to believe the tale and almost dies in his
attempt to retrieve the orchid, whose scent is so powerful (like the "fumes of a
poisonous sickish-sweet drug") that it acts like a narcotic and those who inhale
it cannot live without it; the woman and the orchid are both addictive, potential

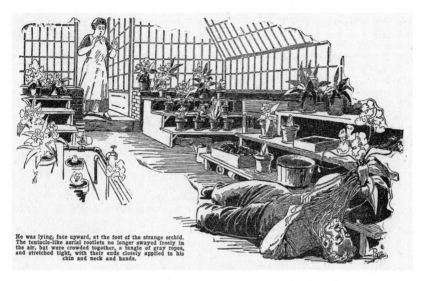

He was lying, face upward, at the foot of the strange orchid. The tentacle-like aerial rootlets no longer swayed freely in the air, but were crowded together, a tangle of gray ropes, and stretched tight, with their ends closely applied to his chin and neck and hands.

Figure 34. Hugo Gernsback often found he didn't have enough new science fiction tales to fill *Amazing Stories,* so he republished H. G. Wells and others; "The Flowering of the Strange Orchid" appeared in the March 1928 issue.
Credit: Collection of George Morgan.

lethal drugs. (No real orchid has narcotic properties — or, if they do, the orchid experts I have consulted are keeping very quiet about it — but in the movie *Adaptation,* one of the numerous images of irresistible desire that is woven around the Ghost Orchid is that it's the source of a powerful drug.) Loring's story ends very enigmatically, leaving the reader wondering exactly what has happened to him and his Goddess.

In stories like the "Orchid Horror" the woman and the orchid merge into a single symbol of evil, an evil so seductive that men may become dependent on it (the Goddess' servant, that "rack-boned skeleton of a man," sounds very like a junkie, desperate to escape his addiction). Similar associations can be found in "The Malignant Flower," by a German author simply identified as "Anthos" (the Greek word for flower).[5] The tale was first published in English in September 1927 by the pulp magazine *Amazing Stories* (edited by Hugo Gernsback, the man who coined the term "science fiction") and concerned another mysterious man-eating flower, but this one is secreted in a hidden valley in the Himalaya, whose location is revealed by a Hindu mystic. The hero, Sir George Armstrong, accompanied only by his faithful servant John Bannister, delays his planned wedding (and his fiancée is not told where he is going, nor why, so that she won't worry). The two men find the hidden valley which is choked with vegetation, including "orchids of the most varied kinds." Pushing through the undergrowth, they find the legendary flower, whose presence is announced—predictably— by an "overcoming strength of scent." As if hypnotized, Armstrong approaches, ignoring Bannister's warning, and

> the flower slowly opened, and something bright and flesh-coloured shot out of it. What darted so suddenly? Was it the sucking arms of an octopus? Was it the soft arms of a woman?

The mysterious tentacles capture Armstrong—Bannister is only just able to cut him out. Armstrong lies on the grass, "a grim and frozen smile as if half of supernatural pleasure, half of the fear of death was on his rigid features." He lives but, as the Hindu had promised, is dead "as far as all earthly love is concerned" (which makes this, to put it mildly, an odd expedition for a man to undertake on the eve of his marriage). He is returned to civilization but never recovers, lingering on in an asylum for 18 months before he dies. A few years later, the bereft fiancée reads about the "Man eating plant" in a newspaper; the species is *Cypripedia* [*sic*] *gigantea*, "belonging to the class of the giant orchids, and the largest flower on earth." (*Amazing Stories* always promoted science facts alongside its fiction, thus helping to blur the boundaries between the two. The magazine accompanied Anthos' story with a short piece on the Titan Arum, *Amorphophallus titanum*, and wondered whether "some bold explorer might venture into still unknown lands and discover a flower even more nearly approximating the description of 'The Malignant Flower'?"[6]) In the story, the orchid is effectively the "other woman" who comes between Armstrong and his bride by seducing him, but there's an odd hint in the story that he has sought this fate, wishing (like some Hindu holy men) to free himself from the passions of the body.

The link between seductive women and orchids was not confined to the pulps, but can also be found in fiction with more serious pretensions, such as *The Woman of Orchids* (1901), by Marvin Hill Dana, an American episcopal priest.[7] The main character is Gilbert Arsdale, a British businessman looking after his mines in the South American Andes while his wife is left at home. He and his secretary are guests at the ranch of a Monsieur Marcou, and Arsdale becomes fascinated with Marcou's wife, whom he first meets exulting at the spectacle of a violent clash between two herds of wild horses (in which Arsdale and his companions are almost killed). "Was it not grand?" asks Madame Marcou and as she is speaking, "an intrepid, ruthless joy shone in her face" (7).

Madame Marcou (she is never given a first name) is a mature, sophisticated beauty, "suave and stately," cultured and highly spirited, who loves riding as much as reading. Arsdale falls for her but is baffled that she is seductive and flirtatious only when the north wind blows; she seems indifferent when it blows from the south. The mystery is solved (sort of) when Madame leads Arsdale, blindfolded, to a secret grove in the forest, which she claims she is the only human who has ever visited before. When she removes the blindfold, he sees an incredible mass of orchids, "the forest vanished, and there was nought save a sea of tremulous fire. As far as the eye could reach was one vast maze of swaying colour."

Arsdale is overwhelmed and briefly forgets his companion until she asks "*Et moi?*"

> The presence of the woman seemed the natural climax of the scene's sorcery. Her eyes met his and her soul leaped in them to his. Without hesitation, tenderly, remorselessly, eagerly, madly, their lips met and their hearts touched." (51)

A little later, Madame explains that the indigenous people call the place "the mount of the demon flowers" and are mortally afraid of it, for reasons that gradually become clear as the air grows still and the "mile of flowers was undulating and writhing as if under a curse of unrest" (as in other stories, no breeze is necessary to set these lively, post-Darwinian flowers in motion). Without explanation, she panics and urges Arsdale to flee with her. As they begin to move, "the vines clung to their impatient bodies" as if trying to detain them and after five minutes of running, Arsdale is desperate to rest, but Madame urges him on. "A sinister desire for slumber benumbed his senses. A strange, sickening odour was in his nostrils and the perfume stupefied him. He longed to sink on the ground and lie there without thought or motion" (56–58).

Finally, they escape the grove's intangible terror and Madame explains what they have fled from:

> Perhaps you know that many orchids are quite odourless: but the most beautiful have a faint scent; and the fairer the flower the more repulsive its perfume.

There where we have been, in the unknown recesses of the forest, is a growth of orchids the most splendid in the world. I do not know its extent. All the space that one can see is filled with their flaming loveliness. They are the most gorgeous, the most fragile blossoms of their kind, and their kind is the fairest of flowers. There those crowded petals hang and quiver on their swaying stems, and there they exhale odours as baleful and deadly as their tints are beautiful. It was the odour of those millions of flowers that made me suspect their location. When the wind blew from the north I approached them closely, as we did to-day, for their scent was carried from me. To-day we were free to bask in their beauty while the wind was in the north; but while we loitered there the wind veered to the east. (62–63)

Since Madame's mood shifts as the wind does, she and the orchids appear mysteriously linked. She apologizes to Arsdale for almost leading him to his death, but he responds that "it was not your fault that the fiery blossoms became as the flaming sword to drive us from our Eden" (64). But perhaps it was: when Madame's daughter Claire and her would-be lover Giles (Arsdale's secretary) accidentally stumble into the same grove of orchids they experience no terror. For them, the wind stays in the north, because the "eagerness and desire and content of the lovers were honest, wholesome, divinely beautiful things, though so modest and simple." They experience nothing but beauty in the grove and they leave the flowers "slowly, reluctantly; but they bore their bliss with them." For the innocent, prelapsarian couple, the orchid grove was indeed a paradise; only the adulterous lovers are punished by the orchids using their intense narcotic scent.

Perhaps the most surprising thing about these stories is that their origin lay in a supposedly true tale of nineteenth-century orchid hunting. In 1896, London's *Daily Mail* published a piece entitled "Most Rare: Flowers That Cost Lives to Secure" that drew together several orchid stories about how men were "continually risking and losing their lives in the attempt to obtain the plants . . . For it is in fever-haunted jungles that the most prized and rarest Orchids are to be found." The author, who signed herself simply "Lady Charlotte," claimed that "a famous Orchid hunter named Fosterman" had been told of a "village of the demon flowers" while exploring in Brazil. (Ignatz F. Förstermann was a real collector, despite the paper misspelling his name, who collected for the orchid king, Sander—although there's no record of him ever going to Brazil.[8])

Förstermann supposedly set off in search of the demon flowers and

> one afternoon, three of his forward guards threw up their arms, and with a
> cry fell senseless to the ground. He had noticed a peculiar sickening odour
> pervading the heavy, heated air, and quickly gave the order for the other
> men to advance with caution and drag back the three fallen ones from the
> spot where they lay. They did so, and returning, reported that they had seen
> through the forest, a little further on, the vast "village of the demon flowers."

Covering his mouth and nose, the collector tried again, but was overcome and finally despaired. A second expedition was mounted, but in vain. "It is a curious fact," as Lady Charlotte noted (in words that surely supplied Dana with the idea for his story), that although "many Orchids are almost scentless, the handsomest ones have a most unbearable smell. When millions of them are collected in a small space this stench, as can easily be imagined, becomes simply intolerable and is literally fatal when long inhaled."[9]

The botanical journal *Orchid Review* reprinted this tale (wryly suggesting it should have been called "The Romance of Orchidology"), and their columnist "Argus" (almost certainly its editor, Robert Allen Rolfe), mocked the *Daily Mail*'s credulous correspondent, noting that the members of the expedition should perhaps try waiting until the orchid's flowering season is over, "in which case I fear its doom is sealed," but perhaps

> that sagacious plant may have heard of Orchid collectors before, and, knowing its weak point, may go on flowering all the year round. I rather hope this will prove to be the case, for such a plant would add a new terror to the Orchid house.[10]

Despite Argus' mocking tone, he nevertheless imagines the orchid to be "sagacious," consciously outwitting the collectors, and acknowledges that it would "add a new terror" to orchid collecting, were such a plant to be found. The imaginary orchids of fiction had seemingly begun affecting even this sober, scientific editor.

Other writers also seem to have been inspired by the story of the "demon flowers." Orchids whose scent was so overpowering they could not be approached appeared in the magazine *Weird Tales* in 1927 in a story called "White Orchids."

And they appeared in a story called "An Orchid of Asia" (1920), which concerns a Venezuelan "Death Orchid" that is collected and hybridized with predictably gruesome results.[11] In stories such as *Woman of Orchids* and "An Orchid of Asia," the flowers' overpowering perfume helps identify them as feminine. In *Woman of Orchids*, after the adulterous lovers escape from the flowers, Arsdale doesn't initially blame Madame for almost causing his death, but his forgiveness does not last; disgusted at his own infidelity, he accuses Madame (and, by implication, her orchids) of leading him astray: "You lull my conscience to sleep, as the poppy would my eyelids, and in that sleep my visions are heavenly, for they are of you; but I must awake, and, like the drugged Oriental, who opens his eyes to a reaction of acute distress, my awakening will be to pain and despair" (87–88). The fiery, animated, and sophisticated woman is turned into a seductive, narcotic, but passive flower (opium often appears in late nineteenth-century fiction as an image of decadent oriental luxury—upon which the supposedly vigorous white races may become dangerously dependent).[12] Wyatt Blassingame's story "Passion Flower" also features an orchid who is both a seductive woman and an irresistible drug.[13] And in "An Orchid of Asia" the Death Orchid is hybridized to create a plant that its creator calls La Revenante (the ghost), whose scent "poured over [the hero in] waves of sensual delight which crippled his will," and under this narcotic influence the hero feels a female presence as if "some monster" were reaching out to him "across aeons of time." The main character is disturbed to find he thinks of the flower "as a living woman" and realizes "he had cultivated a flower beyond the limits of flower life. He had made an orchid-woman . . . a vampire that sat upon his soul."[14]

Women had been metaphorical flowers for centuries, but only in the late nineteenth century did they become flowers that would kill and consume men. The insectivorous plants provided one source for this imaginative transformation. The Venus fly-trap first arrived in Europe in the eighteenth century when John Bartram, a Philadelphia botanist, sent it to his London contact, the Quaker merchant Peter Collinson (who, as we saw in chapter 3, was the first to flower a tropical orchid in Britain). Bartram recounted that the popular name for the plant was "tipitiwitchet," a bawdy Elizabethan term for a vagina.[15] The shape of the plant and its invitingly beautiful, rich red interior make the association almost inevitable, but its toothed edges also recalled the old folk-myth of the *vagina dentata*, which recurs in many cultures—perhaps as a deterrent to would-be rapists—and clearly embodies a fear of castration. It was the association of carnivorous plants with dangerously seductive women that prompted Swinburne's "The Sundew" (1862, see chapter 6).[16] Thanks in large part to Darwin, carnivorous plants and orchids had become linked in the public's imagination in the late nineteenth century, which probably helped orchids become predatory. (The link between popular understandings of Darwin and these tales is fairly explicit in "An Orchid of Asia"; not only is the Death Orchid hybridized with *Angraecum sesquipedale*, the species about which Darwin made his famous moth prediction, but also the collector tells his gardener to "push this intercrossing to the highest limit! Choose the largest, strongest specimens."[17])

However, the ways in which the imaginary orchids began to be imagined as female, as well as more aggressive and more sexual, reflect late-Victorian shifts in the relations between the human sexes. The 1860s marked the first organized campaigns for women's rights in Britain, when concerted attempts were made to get women the vote. The philosopher and member of Parliament John Stuart Mill attempted to amend the 1867 Reform Bill so that a voter would be defined as a "person," not as a "man." The vote was lost, but Mill made a brilliant speech in which he argued that "a silent domestic revolution" was already underway in relations between men and women.[18] Many welcomed this shift towards greater equality, but others felt threatened by the increasing visibility of well-educated, self-confident women and their demands for a greater role in society. It is surely no coincidence that the French phrase *"femme fatale"* entered the English language at around the same time; in 1879, an American newspaper used it to describe "a woman whose influence brings a curse to all within its range," a description that would clearly apply to Madame Marcou or the nameless Goddess of the Orchids.[19]

Orchid stories were, of course, only one place where the *femme fatale* made her debut in the late-Victorian imagination; she expressed a range of anxieties about female sexuality and other threats to manly imperialism.[20] In the 1860s, women's votes were the main issue, but by the 1890s (when the word "feminist" first entered the English language) a few women were campaigning around explicitly sexual matters and these were reflected in a new fictional type known as the "New Woman." Grant Allen, who did so much to publicize Darwin's "queer flowers," was best known for his novel *The Woman Who Did* (1895), which examined the dangerous sexual freedoms being demanded by the New Woman of the *fin de siècle*.[21] The phrase "New Woman" described those, like Allen's heroine, who believed in free love, but increasingly referred to all the young, single women who chose to work instead of marrying, thus giving themselves a degree of freedom. Most New Woman novels were written by women, many of whom were also campaigners for women's rights, and their novels tended to focus on the double standards that often accompanied discussions of male and female sexual behavior. The fact that married women had no right to refuse their husband's sexual demands led some campaigners to describe marriage as little more than legalized prostitution. And some women focused their energies on such unladylike topics as contraception, sex education, and the spread of venereal diseases. When a young woman named Edith Lanchester (a friend of Karl Marx's daughter, Eleanor) decided to refuse marriage and simply live openly with the man she loved, her father and brothers decided she must be mad; they kidnapped her and imprisoned her in a private asylum. Women rallied to her cause and she was freed, but even her supporters didn't entirely endorse her decision; the feminist magazine *The Woman's Signal* suggested she had been led astray by reading Allen's *The Woman Who Did*.[22]

Women's demands for sexual as well as political equality disturbed many people (both men and women), and these anxieties were often apparent in the fiction of the period, including the orchid stories. When writers address sex and reproduction, they often choose (or are forced) to confront questions about what (if anything) about human behavior is "natural," questions that may force us to confront our prejudices. For example, despite Allen's support for progressive causes, including feminism, his view of women was colored by his interest in biology; he, like many progressive men of the period (and perhaps since), found it hard to disentangle his liberal political convictions from deeply held preconceptions about women's supposedly "natural" role as mothers.[23] Although he believed strongly in women's rights to independence and self-expression, Allen seems to have worried that women might simply lose interest in being wives and mothers once they had a wider range of options available to them. What would become of society? And where would he find a secretary to type his articles for him?

Predatory orchids who became female and predatory women who became orchids hint at the discomfort some felt when women began to demand a change in their supposedly natural roles. However, the worries aroused by the carnivorous orchid are always intermingled with desire; orchids, like *femmes fatales*, are so dangerous precisely because they are so alluring. The same potent mix of dread and desire is apparent in the imperial anxiety that lurks in the orchid stories, a fear of the dangerous tropics, ripe with sickness and scheming natives, embodied in seductive exotic women (more seductive and exotic than their more civilized sisters back home, perhaps?). Wells' "Strange Orchid" seems to mock orchid collectors like Mr. Wedderburn, whose lust for flowers masked their fear of women. Wedderburn yearns for some excitement in his dull bachelor life (one of the things he envies about the dead orchid-collector, Batten, is that he was married twice), but the orchid's sensuality almost kills him (and it takes a quick-thinking, but apparently chaste, woman to save his life). Such dangerous orchids seem to imbue women with qualities that were simultaneously repellent and seductive (precisely the way in which a few people regard orchids themselves); could such women be enjoyed and domesticated, like the flowers? The potent danger of the sexually assertive woman was hinted at by Arnold Bennett in his novel *The Pretty Lady* (1918), in which the narrator imagined the courtesan he had fallen for becoming his kept woman, so that "her temperament extreme" could be safely available at his pleasure. Given the right environment "she would blossom," and "she would become another being, that was to say, the same being, but orchidised."[24] For many writers, however, the seductive, exotic orchid was a dangerous trap (specifically a man-trap).

Fiction's lethal orchid women suggested the ways in which long-standing ideals of femininity were breaking down. But precisely because orchids could symbolize the breakdown of traditional families and morals, they were celebrated by the fin-de-siècle movement known as the Decadents, for whom an orchid could embody alternatives to conventional forms of both masculinity and sexuality. Among the Decadents' leading British figures was Oscar Wilde, who vigorously attacked the use of "exotic" as a term of abuse. In his book *The Soul of Man under Socialism* (1891), Wilde described the contempt for the exotic as "the rage of the momentary mushroom against the immortal, entrancing, and exquisitely lovely orchid."[25]

Wilde and many of his friends were attracted to orchids by exactly those qualities that others found repulsive in them: their apparent artificiality and exotic grotesqueness. In Wilde's novel, *The Picture of Dorian Gray* (1890), the eponymous hero loves orchids and his friend and mentor, Lord Henry Wotton, uses them to explain his dissatisfaction with the world around him:

> Yesterday I cut an orchid, for my button-hole. It was a marvellous spotted
> thing, as effective as the seven deadly sins. In a thoughtless moment I asked

one of the gardeners what it was called. He told me it was a fine specimen of *Robinsoniana*, or something dreadful of that kind. It is a sad truth, but we have lost the faculty of giving lovely names to things. Names are everything. I never quarrel with actions. My one quarrel is with words. That is the reason I hate vulgar realism in literature (212–13).[26]

This languid aesthete encourages his young protégé, Dorian Gray, to adopt his philosophy of life and—in order to inspire him—presents him with a nameless yellow novel written in a "curious jewelled style, vivid and obscure at once . . . There were in it metaphors as monstrous as orchids and as subtle in colour" (139). This plotless and "poisonous" book inspires Dorian to emulate its fictional hero, "a certain young Parisian who spent his life trying to realize in the nineteenth century all the passions and modes of thought that belonged to every century except his own." Inspired by this example, Dorian tries on various intellectual fashions, including the "materialistic doctrines of the *Darwinismus* movement" (inspired by Darwin's German follower, Ernst Haeckel). While there's no evidence that he takes any interest in orchid biology, he does decide that the best way to celebrate committing a murder, and successfully disposing of the body, is to order a huge number of orchids.

The disgraceful yellow book that served as Dorian's unholy bible was modelled on a real novel, *Against Nature* (*À Rebours*, 1884) by the French writer Joris-Karl Huysmans.[27] The hero of Huysmans' novel is Jean des Esseintes, who has always "despised the common, everyday varieties that blossom on the Paris market-stalls." Rather like the English orchid collector, James Bateman, des Esseintes saw in the florist's window "a microcosm in which every social category and class was represented." Flowers like carnations were "poor, vulgar slum-flowers," that were "really only at home on the window-sill of a garret." The middle classes were exemplified by "stupid flowers such as the rose," which he described as "pretentious and conventional," whose proper place was "in pots concealed inside porcelain vases painted by nice young ladies." The highest class were represented by "exotic flowers exiled to Paris and kept warm in palaces of glass," which lived "aloof and apart, having nothing whatever in common with the popular plants or the bourgeois blooms" (82).

Predictably, des Esseintes is drawn to the delicate exotics, "rare and aristocratic plants from distant lands, kept alive with cunning attention in artificial

VIII.

IL avait toujours raffolé des fleurs, mais cette passion qui,
pendant ses séjours à Jutigny, s'était tout d'abord étendue à
la fleur, sans distinction ni d'espèces ni de genres, avait fini
par s'épurer, par se préciser sur une seule caste.

Depuis longtemps déjà, il méprisait la vulgaire plante qui s'épa-
nouit sur les éventaires des marchés parisiens, dans des pots
mouillés, sous de vertes bannes ou sous de rougeâtres parasols.

En même temps que ses goûts littéraires, que ses préoccupations
d'art, s'étaient affinés, ne s'attachant plus qu'aux œuvres triées à
l'étamine, distillées par des cerveaux tourmentés et subtils ; en
même temps aussi que sa lassitude des idées répandues s'était
affirmée, son affection pour les fleurs s'était dégagée de tout
résidu, de toute lie, s'était clarifiée, en quelque sorte, rectifiée.

Il assimilait volontiers le magasin d'un horticulteur à un micro-
cosme où étaient représentées toutes les catégories de la société :

Figure 35. Decadents like des Esseintes were drawn to orchids,
"rare and aristocratic plants from distant lands, kept alive with cunning
attention in artificial tropics created by carefully regulated stoves."
(Illustration by Auguste Lepère, from J.-K. Huysmans, *À Rebours*, 1903)
Credit: Bibliothèque nationale de France.

tropics created by carefully regulated stoves"; their fragile vulnerability, especially their dependence on artificial heat, is evidence of their aristocratic bear-
ing. He begins to fill his home with artificial flowers, but soon replaces these
with wagonloads of exotics, especially "natural flowers that looked like fakes."
Orchids are among his most treasured finds:

> The *Cypripedium*, with its complex, incoherent contours devised by some
> demented draughtsman. It looked rather like a clog or a tidy, and on top was a
> human tongue bent back with the string stretched tight, just as you may see it
> depicted in the plates of medical works dealing with diseases of the throat and
> mouth; two little wings, of a jujube red, which might almost have been bor-
> rowed from a child's toy windmill, completed the baroque combination of the
> underside of a tongue, the colour of wine lees and slate, a glossy pocket-case
> with a lining that oozed drops of viscous paste.

Understandably, he found he "could not take his eyes off this unlikely-looking
orchid from India" (85).

In des Esseintes' post-Darwinian world, the omniscient creator has become
"a demented draughtsman," the wonderful contrivances look like children's
toys, and the overall effect is sinister, evoking medical illustrations of diseased
mouths, oozing mouths nobody would want to kiss (other flowers remind him
of hereditary syphilis). His floral experiments lead him to precisely the opposite
conclusion that Darwin reached; the Englishman had been awed by the slow
power of nature's selection when compared with that of human plant or animal
breeders. By contrast, when des Esseintes reflects on his orchids:

> "There's no denying it," he concluded; "in the course of a few years man can
> operate a selection which easy-going Nature could not conceivably make in
> less than a few centuries; without the shadow of a doubt, the horticulturalists
> are the only true artists left to us nowadays." (88)

What is striking—and strikingly decadent—about des Esseintes' conclusion is
that it's the very artificiality of the horticulturalist's work (for which Darwin
coined the term "artificial selection") that makes the results deserve the term of
art; nature is only interesting when it imitates art.

Another of des Esseintes' orchids, a *Cattleya*, has "a smell of varnished deal, a toy-box smell that brought back horrid memories of New Year's Day when he was a child" (87). A smell that evokes a childhood memory is reminiscent of Marcel Proust, for whom the taste of a Petit Madeleine has the same effect, returning him to a precise moment in his childhood. The main character of Proust's *In Search of Lost Time* (*À la Recherche du Temps Perdu*), Swann, has a mistress, Odette de Crécy, who explains why orchids — especially *Cattleyas* — are among her favorite flowers in terms that are strikingly similar to those used by des Esseintes:

> They had the supreme merit of not looking in the least like other flowers, but of being made, apparently, out of scraps of silk or satin. "It looks just as though it had been cut out of the lining of my cloak," she said to Swann, pointing to an orchid. (265)[28]

The decadent sensibility rejoiced in such paradoxes as real flowers that looked fake and Proust describes the orchid as Odette's "sister" — as if the woman were another richly artificial flower, an orchid in a room full of highly decorated, oriental ornaments. The first time Swann and Odette touch, the *Cattleyas* provide the pretext; he arranges a bouquet of them "in the cleft of her low-necked bodice," and a delicate brushing away of spilt pollen ends in "her complete surrender." At each subsequent meeting, the *Cattleyas* provide a pretext for intimacy to the point where the lovers use the phrase *"faire Cattleya"* (which is really untranslatable, but can be crudely rendered as "do a Cattleya") as a euphemism for making love (279–281).

Boy's Own Orchids

Despite, or perhaps because of, the rather unsavory reputation of orchids, various Christian writers made strenuous efforts to rescue them from the effects of Darwin's "flank movement." The Reverend James Neil published *Rays from the Realms of Nature: Or, Parables of Plant Life*, in which he took a series of virtues and shoehorned various plants into the role of illustrating them.[29] He even attempted to cleanse the testicle-like double bulbs of the Early Purple orchid from a 2,000-year-old association with sex, by using the flower to illustrate the virtue of "Growth with Grace." According to Neil, the new bulb that the orchid grows each year always appears on the southern side of the plant:

Thus it may be observed steadily travelling on to the bright home of this family
of flowers in the tropics — the cloudless land of sun. Just so the soul that has
haven for its home patiently grows heavenward, and each year throws out
thither, as it were, new roots of holy affections.

No hint of steamy sexuality in Neil's vision of the tropics! His natural history
is even shakier than his theology; he suggests that orchids were spontaneously
generated, since they "seem to be called into existence by a warm humid atmosphere and decaying vegetable matter," and suggests that perhaps that is why
they migrate to the tropics, where these processes are "continually proceeding
on so rapid a scale" (67). Tropical epiphytes are used to illustrate the principle
of charity, because their aerial roots absorb the "foul and poisonous gases" of
the jungle air, transforming them "into the perfume of its own sweet flowers"
(25–27).[30]

In similar vein, orchids featured regularly in the wholesome, morally improving magazine for future empire builders, the *Boy's Own Paper*. The magazine
had been founded by an evangelical Christian publishing organization, the Religious Tract Society, to counter the growing popularity of cheap, secular weeklies
known as "penny dreadfuls." The paper regularly recommended orchid collecting as an improving hobby for young men. In the spring of 1879, for example, the
author of "The Boy's Own Flower Garden, etc." told his readers that fresh air and
exercise were not enough to justify a country walk, it is also "a good thing to have
an object for a walk," and for those who lack a suitable goal, "I recommend you,
just now, to go out orchis hunting" and proceeds to describe the various native
kinds. Apart from a brief acknowledgment of "the patient researches of Dr. [*sic*]
Darwin and others," no hint was given that anything about orchids might be unsuitable for impressionable young minds. Indeed, the writer piously trusts that
studying insect fertilization will lead young men to "lift up our hearts in thank-
fulness to the Great Author of all."

Paradoxically, one reason Christian writers tried to rehabilitate orchids and
their collectors was that, despite ruling over the largest empire the world had
ever seen, by the end of the nineteenth century the British were becoming preoccupied with imperial decline. Were they still the vigorous manly race who
had subdued half the world? Were they strong enough to hold on to what they
had won? Despite their popularity, orchids retained their association with luxury, wealth, and artificiality. Writers often used them as symbols of the idle (and

inbred) rich—feeble and unnecessary. The Anglo-Irish playwright Edward John Moreton Drax Plunkett, eighteenth Baron Dunsany, was an orchid collector himself and he used rich, purple orchids as an image of decadent luxury in his play *The Laughter of the Gods* (1917), which dealt with the decadence of ancient Babylon.[31] The Babylonian king Karnos leaves his city to see the fabulous orchids of Thek, ignoring the vengeful gods who are destroying his kingdom. "It is like a picture done by a dying painter," he says of the orchid-filled jungle around Thek, "full of a beautiful colour. Even if all these orchids died to-night yet their beauty is an indestructible memory."[32]

Plunkett's image of a decaying civilization is typical of a widespread anxiety over what was often called Degeneration, a fear that the sun was setting over the great European empires that was partly prompted by Darwinism. Darwin's ideas were commonly interpreted as proving that competition (between species, individuals, companies, and nations) was a law of nature, and one that (for those who accepted the idea of a godless universe) was humanity's only guarantee of progress. That promise of progress offered some consolation for the loss of natural theology's reassurances, and helps explain why Darwin's ideas spread so rapidly in Britain. However, as late Victorians came to accept Darwin's message, many were disturbed by the thought that modern civilization was making life too easy. What would happen as our civilized compassion led us to protect the weaker members of our species from the rigors of natural selection? What would happen once we stopped evolving? (That was precisely the question Wells asked in *The Time Machine*.) As the nineteenth century became the twentieth, orchids represented a fear of the corrupting influence of foreign luxury (around the turn of the nineteenth century, the English sometimes used terms like "orchidaceous" and "showy" to express contempt for "exotic" foreigners). The writer Richard Le Gallienne (who, despite his name, was born in Liverpool), a friend of Oscar Wilde, bemoaned the complexity of modern life, and arguing that "we are overbred. The simple old type of manhood is lost long since in endless orchidaceous variation."[33] Le Gallienne's words reveal a faint but pervasive fear that humanity, like the hothouse orchid, was becoming weak because we had become so dependent on the artificial luxuries of modern civilization, and thus increasingly unable to survive in the harsh world of nature.

The fear of imperial decay, of the degeneration of Britain's imperial strength, helped to shape a number of popular stories that tried to rescue orchids and their collectors from the unsavory clutches of both orchidaceous seductresses and decadent orchid fanciers. In 1892 (just a year after the rediscovery of *Cattleya labiata*), the *Boy's Own Paper* ran a serial that would later be published as a book called *The Orchid Seekers: A Tale of Adventure in Borneo*, which took the search for a mysterious blue orchid as the excuse for various types of stereotypical, imperial derring-do.[34] The main character is Ralph Rider (based on Frederick Sander), "a squarely built, broad-shouldered man, in the prime of his life, with a ruddy, kind countenance; his hair and beard were glossy brown, with not a grey hair in his beard." The story is set in 1856, because "botanists and amateurs who recall that time think of it as a golden age!" Orchid houses were still rare and vast tracts of the earth lay unexplored and unexploited; "an orchid grower in those days had all the world before him" because the "tropic realms" were measureless spaces "where imagination roamed unchecked" (1–2).

The tale opens with Rider in his orchid house, in conversation with a younger man, who is similarly macho:

> He stood six feet two in his stockings. His oval face was of the colour of burnished copper, bronzed by the scorching tropical sun, which had bleached the heavy flaxen moustache at least two shades lighter than his hair, but had not dimmed the lustre of the blue, watchful eyes.
>
> His nationality could be seen at a glance almost—he was a German. His name, Ludwig Hertz. Occupation, a collector of orchids. He had only one arm, the left. In the place of the right, lost in some machinery, he had an iron hook, as useful as some people's fingers. Indeed, his dexterity with that curved piece of iron was something to marvel at. (2)

The piratical hook sounds like an invention too far, but of course it identifies the real-life model for the fictitious Hertz; he is clearly based in part on Benedict Roezl. The similarity between the real Roezl and the fictitious Hertz is hardly surprising—Frederick Boyle was coauthor of *The Orchid Seekers*. The story's main author, Ashmore Russan, had contacted Sander of St. Albans for information about orchids. They had put him in touch with Boyle, who—as we have seen—acted as an unofficial, one-man PR department for the firm and was always reluctant to let the facts get in the way of a good story.[35] According to the editors of the *Boy's Own Paper*, Boyle agreed to consult, and the "outline of the tale is his." It describes "scenes he had himself beheld, of peoples and individuals he himself had known," so he added a few sentences here and there to

THE BOY'S OWN PAPER

No. 681.—Vol. XIV.   SATURDAY, JANUARY 30, 1892.   Price One Penny.
[ALL RIGHTS RESERVED.]

THE ORCHID SEEKERS:

A STORY OF ADVENTURE AND PERIL
IN BORNEO.

By Ashmore Russan and Frederick Boyle.

CHAPTER I.—A PALACE OF ENCHANTMENT.

A WELL - FILLED and well - cared for orchid house, when the plants are in blossom, is indeed

INTRODUCTORY NOTE.

Some of our boy readers may cry, on seeing the names of two authors at the head of this page, "What! A couple of 'em?" And at any point they will be likely to ask themselves, "Now, I wonder which of the two wrote this?"

The public in such cases is commonly left to wonder, but circumstances here are unusual. When Mr. Ashmore Russan formed a project of writing a story upon the subject of orchid-collecting, he applied to Messrs. Sander, of St. Albans, the great importers of orchids, for special information. They referred him to Mr. Frederick Boyle, as one who had travelled in many lands where those plants flourish, who grows them, as an amateur, with unusual success, and publishes much about them. Mr. Boyle had no time to take an equal part in writing the story, but he consented to advise, direct, and in general to lend his assistance. The outline of the tale is his, and for all statements therein, historical, local, or scientific, Mr. Boyle is responsible. It naturally happened, since he was treating of scenes he had himself beheld, of peoples and individuals he himself had known, that he found it necessary to take the pen from Mr. Russan's hand and write a few lines here and there. But in general he confined himself to his functions of director and critic. It will be understood, after this explanation, that the tale rests on a solid basis of fact all through.

"'Ach! I am robbed.'"

Figure 36. The one-armed Ludwig Hertz appalled to find one of his
newly discovered orchids already growing in Ralph Rider's greenhouse.
(From the first installment of Ashmore Russan and
Frederick Boyle, *The Orchid Seekers,* 1892)

describe these, but the bulk of the work was Russan's. "It will be understood," the editors added, "after this explanation, that the tale rests on a solid basis of fact all through."

Russan and Boyle decided to make their fictitious orchid collector a German, so perhaps he was based in part on Micholitz too. Their decision gave them the opportunity to have the fictitious Hertz speak—for almost 400 pages—in what can only be described as "Music Hall Kraut." When asked by Rider if he thinks the fabulous blue orchid really exists, Hertz replies: "Vhy nodt? . . . Nodings ish too marfellous. I have seen so many marfels dat I gan beliefe anydings" (5). It is no coincidence that the flower at the center of the quest should be blue, since that is probably the rarest color for an orchid; in 1896 the *Daily Mail* told its readers that "the list of these is short indeed, even when those which exist only in the tales of Orchid hunters are taken into account."[36] Boyle recalled that in his early years in Borneo, he had heard tales of a blue orchid that had been sent to London, but had no idea what species it was ("we were desperately vague about names in the jungle at that day") nor whether it survived in cultivation.[37]

In Russan and Boyle's story, an expedition is launched to find the flower and Rider's teenage sons, Jack and Harry, are sent with Hertz to explore the jungle. He warns them of the many dangers, but Jack replies:

> "When a fellow has made up his mind not to stay at home if he can help it; when he wants to travel and find orchids, and feels that he'll do no good at home, what then?"
>
> "He had besser gome mit me."
>
> "Hurrah!"

There is no doubt that Hertz is qualified for such a journey. We are told that "had he been dropped from a balloon in any part of the Tropics an examination of the flora would have told him his locality." Like Wells' fictitious collector, Batten, Hertz has already had many adventures, including "escapes from savage animals and equally savage men, from miasma-breathing swamp, river flood, and ocean storm," but these were "all in the way of business. He loved his occupation, and would not have exchanged it for a dukedom" (3). But what about the boys? Why is their father convinced that they are capable of similar heroism, while he stays home to stop their mother worrying? Well, Jack, the elder, was "fresh coloured and youthful . . . but very determined." More importantly, he "could jump, with a run, five feet five in height; he could throw a cricket ball one hundred and ten yards," which presumably equipped him either to whack orchids out of high trees or knock treacherous natives for six. His younger brother was "much more book-worm," but "it would have been impossible for Ralph Rider's sons to belong to the 'cling-mother's-apron-string' species of the genus 'Molly Coddle'" (19). Hurrah indeed.

Clearly the writers, and the editor's *Boy's Own Paper*, were anxious to persuade their young readers that there was nothing effeminate about orchid collecting. The phrase "Mollycoddle" derives from the eighteenth-century English slang, "molly" meaning a homosexual or effeminate man; the dubious associations that orchids were acquiring in the 1890s seemed to have filled the story's writers with a nostalgia for the clean, wholesome 1850s, when men were men, orchids were orchids, and nobody had heard the name Oscar Wilde (sentenced to penal servitude in 1895, just two years before *The Orchid Seekers* was published as a book).

The writers continue to protest too much about the hyperbolic masculinity of their heroes when they reach the steaming, cliché-infested jungles of Borneo, where neither the writing nor its sensibilities improve. The swarming Malay pirates are short, dark, and cowardly, the Chinese are teeming and inscrutable, the Greeks are the world's "dirtiest scamps," and in general those whom Hertz calls "natifes" are untrustworthy, violent headhunters (or possibly worse; all very reminiscent of Micholitz's comments, see chapter 6). Thankfully, the flow of offensive stereotypes is broken up by interesting information about orchids, many of which are described accurately as Hertz lectures the boys on the definition of an orchid, gives the scientific names of many, and even describes their intimate relationships with their pollinators. Setting the book in 1856 ensures that Darwin's name never needs to be mentioned, which must have been a relief to the story's evangelical publisher, yet much of the information about orchid pollination has clearly been derived from his work. (Even *Angraecum sesquipedale* and its predicted pollinating moth get a mention.) Hertz describes the way *Coryanthes* traps a bee, "vets his vings" in its bowl-shaped labellum full of liquid, so that it cannot fly away but must crawl through the flower and pick up a load of pollen in the process; Darwin's specter seems to haunt the story, but Hertz has a different explanation of the wonders of orchids. He tells Jack and Harry, "Nature ish vonderful—vonderful! Bear in mind, boys, vhen I say 'Nature' I mean 'Gott!'" (131).

Despite their muscular Christianity and ingrained racial superiority, our intrepid heroes fail to collect the prized blue orchid. However, the authors insist the story is true and their tale doubtless inspired many of their young readers to study botany (and practice cricket) more diligently than ever, in the hope of hunting their own orchids in far-off jungles. The book did well enough to prompt a sequel, *The Riders; or, through Forest and Savannah with the Red Cockades*.[38] And Russan and Boyle's books were not the only example of the orchid-hunting genre; a few years earlier Percy Ainslie had published *The Priceless Orchid: A Story of Adventure in the Forests of Yucatan* (1892), featuring another young orchid-hunter called Jack sent off to the wilds to find the incredibly rare *Cattleya dolosa* and thus complete a rich orchidmaniac's collection. It was described by a reviewer as "a decidedly interesting story" that "can hardly fail to interest boys."[39]

Africa provided the setting for another orchid hunting tale, by the celebrated H. Rider Haggard, who promoted the manly ideals of empire through adventures such as *King Solomon's Mines* and *She*. In *Allan and the Holy Flower* (1915), Haggard (who was a keen orchid-grower himself) embroiled his great white hunter, Allan Quatermain, in a hunt for "the most wonderful orchid in the whole world," in a tale of old-fashioned heroism that must have offered many of its readers a welcome distraction from the developing horror of the First World War.[40] In the story, Quatermain takes the only example of the flower, a single dried specimen to London, to an orchid sale at an auction house looking for someone to finance an expedition to bring back living specimens that would be worth a fortune. At the auction, he is warned that "there are people in this room, Mr. Quatermain, who would murder you and throw your body into the Thames for that flower" (34). The warning comes from a rather feckless young man, Stephen Somers, whose extravagant love of orchids disgusts his wealthy father: "You don't even spend your money, or rather my money, upon any gentleman-like vice, like horse-racing or cards or even—well, never mind," complains Stephen's father, "you take to flowers, miserable, beastly flowers" (46).[41] Clearly, even wasting his father's money on women would be better than on effeminate orchids. Cut off without a penny, Stephen goes to Africa with Quatermain, who warns him that they will probably find nothing but "a nameless grave in some fever-haunted swamp," but for Stephen the danger only adds to the attraction. After adventures with slave traders, mad naturalists, and a tribe of "good" (by which Haggard always meant "subservient") Africans, they dis-

Figure 37. The perils and rewards of orchid-hunting inspired
a number of similar fictions in the late nineteenth century.
(From Percy Ainslie, *The Priceless Orchid*, 1892)

Figure 38. Young Stephen Somers and Allan Quatermain
get their first sight of "the most wonderful orchid in the
whole world." (Illustration by Maurice Greiffenhagen, from
H. Rider Haggard, *Allan and the Holy Flower*, 1915)

cover that the fabulous orchid is worshipped by an evil tribe of cannibals, whose ape god (an albino precursor to King Kong) is eventually slain by Quatermain with a single bullet. Stephen grows in courage and stature during the adventure, ending up with not just the orchid, but also the love of a good woman (a kidnapped white woman he helps Quatermain to rescue from the cannibals), and even his father's forgiveness. Clearly Russan and Boyle were not the only ones who thought orchids would nourish "that longing to roam which is the heritage of British boys."[42] These stories seem to propose that — as long as they remained great and white — hunters could pursue flowers as readily as lions and not compromise their manliness.

Real and fictional orchid hunters struggled with each other, with wild animals, tropical diseases, and occasionally "natifes" as they sought to gain control over these precious flowers. Meanwhile, nineteenth-century writers were locked in a parallel struggle as each offered competing definitions of the orchid's symbolic significance. Some writers celebrated the supposed artificiality of orchids, while others insisted that to study orchids was to examine nature, pure and untrammeled. Yet whether a piece of writing about orchids was sober, scientific, entertaining, exciting, moralizing, or morbid, every writer played a role in transforming orchids into artifacts, entities no longer "natural" but shaped by human culture. The more people wrote about orchids, the more varied and complex their meanings became. In the hands of twentieth-century writers, the orchid's associations with sex, death, and luxury would take them from the fictional jungles of empire to the modern urban jungle, a setting in which the figure of the manly white hunter would be replaced by a new embodiment of beleaguered masculinity, the private eye. In the modern city, cannibals were replaced by gangsters, the worship of pagan gods gave way to that of sex and money, but the orchids still flourished.

Plate 11. Flowers of evil: artist Tom Adams combined orchids and carnivorous flowers in a new cover for Raymond Chandler's classic *The Big Sleep* (Ballantine Books, 1971). Credit: Tom Adams.

Plate 12. A digger wasp (*Argogorytes mystaceus*) attempting to copulate with a fly orchid (*Ophrys insectifera*) at Downe Bank, Kent, in June 2005. Darwin collected orchids at the same site 150 years earlier, but never understood how species like this were pollinated. Credit: Grant Hazlehurst.

Plate 13. Cattleya mendelii, a South American orchid for which Albert Millican was willing to cut down over four thousand trees. (From *Travels and Adventures of an Orchid Hunter,* 1891).

Plate 14. Odontoglossum crispum, a variety of which fetched 650 guineas (roughly £298,000/$471,000 today) for a plant that "consisted of one old bulb and one fine new bulb with a leaf eight inches long." (From F. Sander and Co., *Reichenbachia: Orchids Illustrated and Described*, 1888)

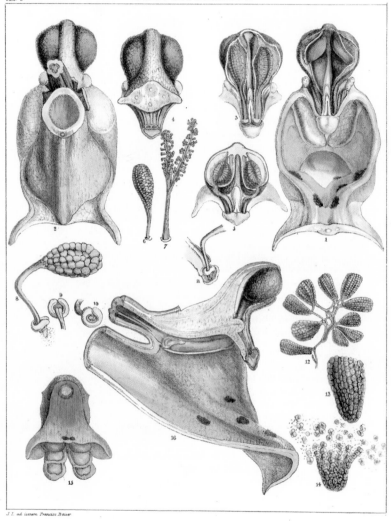

J. L. ad. vivum. Francisc Bauer

Plate 15. Nineteenth-century classification methods required dissecting flowers
and the resultant illustrations make the orchids look very alien indeed.
Illustration by Franz Andreas Bauer.
(From John Lindley, *Illustrations of Orchidaceous Plants*, 1830–38)

9

Manly Orchids

Inside the greenhouse, "the air was thick, wet, steamy and larded with the cloying smell of tropical orchids in bloom. The glass walls and roof were heavily misted and big drops of moisture splashed down on the plants. The light had an unreal greenish colour, like light filtered through an aquarium tank. The plants filled the place, a forest of them, with nasty meaty leaves and stalks like the newly washed fingers of dead men. They smelled as overpowering as boiling alcohol under a blanket."[1] And into the orchid house comes a man who is not himself mean, who is neither tarnished nor afraid: Philip Marlowe, private eye, the hero of Raymond Chandler's first novel, *The Big Sleep* (1939).

The "cloying smell" of Chandler's orchid house is the unmistakable reek not merely of sex, but of sex corrupted, gone bad. When Marlowe's client, General Sternwood, asks him if he likes orchids, Marlowe replies "not particularly." The General, despite owning a greenhouse full of them, concurs: "They are nasty things. Their flesh is too much like the flesh of men. And their perfume has the rotten sweetness of a prostitute." (In the 1946 Howard Hawks movie, starring Humphrey Bogart and Lauren Bacall, this had to be changed to "the rotten sweetness of corruption" to avoid censorship.) The orchids' "nasty meaty leaves" and their resemblance to dead men's fingers hint at the murders to be com-

Figure 39. General Sternwood (Charles Waldron) explaining to
Philip Marlowe (Humphrey Bogart) that orchids "are nasty things."
(From *The Big Sleep*, directed by Howard Hawks, 1946)

mitted and discovered later in the book; the second image is drawn—as we have
seen—from Ophelia's death scene in *Hamlet*, but in Chandler's world, women
are more likely to be murderers than victims.

Why would a man who dislikes orchids spend a fortune on growing them?
The aging general explains to Marlowe that "I seem to exist largely on heat,
like a newborn spider, and the orchids are an excuse for the heat." The general's characterization of their scent, its "rotten sweetness," connects the flowers
with his own wayward daughters, neither of whom, he casually notes, "has any
more moral sense than a cat. Neither have I. No Sternwood ever had." He assumes his daughters have "all the usual vices" and their illegal pastimes have led
to his being blackmailed ("again," he wearily acknowledges), which is why he
has contacted Marlowe. The orchids serve as metaphors for Sternwood's daughters, particularly the younger one, Carmen, who likes "the flesh of men" rather
"too much" and is conspicuously perfumed, sweet, and rotten. The link between
orchids and sex is maintained when later in the novel, Marlowe picks up an obscene volume that has been dropped by the nervous client of the blackmailing
pornographer he's pursuing; as he flicks through it, the detective notes that "the
air was as still as the air in General Sternwood's orchid house." The General's
explanation of the orchid collection also carries the faint implication that his
daughters were not the result of any desire to continue his family's name, but
merely an excuse for an old man to exercise his lust—the old general's "heat."

Women, his daughters and their mother, disgust the old general in the same way that the orchids do, they were just an excuse for the heat.[2]

Some have found the orchid-house scene in *The Big Sleep* rather puzzling, because the orchids are never really mentioned again and the General becomes increasingly marginal to the story.[3] However, for readers of 1930s crime fiction, the presence of orchids would immediately have evoked one of the most successful and famous fictional detectives of the day, Nero Wolfe.

Wolfe had been created by Rex Stout five years earlier in 1934, and would star in more than 30 novels and many more short stories. He was enormously fat (weighing 300 lbs., or 137 kg) and enormously wealthy, he used his extraordinary intelligence to solve crimes merely to finance an extravagant hobby: orchid growing. His sidekick, Archie Goodwin, was Wolfe's Dr. Watson and he narrated their adventures. Archie recounts, "Wolfe had once remarked to me that the orchids were his concubines: insipid, expensive, parasitic and temperamental. He brought them, in their diverse forms and colors, to the limits of their perfection, and then gave them away; he had never sold one."[4] (Wolfe is clearly a literary descendant of des Esseintes and Dorian Gray.)

The orchids (10,000 of them) lived in a purpose-built rooftop greenhouse above Wolfe's large New York brownstone on the upper West Side and he maintained an inflexible routine of tending them with the aid of his gardener, Horstmann; mere crimes were never allowed to interrupt his orchid routine. Fastidious in everything, immaculately dressed (beginning the day with breakfast in canary-yellow pajamas and matching slippers with turned-up toes), he was so fat he could hardly leave his own home (Archie does all the legwork for his cases). The crimes he solves are as preposterous as his character. Everything in these books is overheated and ridiculous; Stout simply took Sherlock Holmes, transplanted him to Manhattan, and then tried to out-Sherlock Sherlock at every turn: Holmes' mild distrust of women becomes Wolfe's outright misogyny; the London detective's violin playing becomes the New Yorker's conviction that "all music is a vestige of barbarism" (*Blood Will Tell*, chapter 2); and, since Holmes occasionally indulges in a 7% solution of cocaine when he has no fascinating cases to occupy his mind, Wolfe had to have his own uncontrollable addiction, which crime-fighting allows him to feed—orchids. The plants are nothing but a pretext for a series of preposterous plots; in *Some Buried Caesar* (1939), for example, which was apparently Stout's favorite, the story opens with Wolfe and

Archie driving over 230 miles to enter an orchid show because Wolfe "desired and intended to make a monkey" of Charles E. Shanks, a rival breeder "who has twice refused to trade crossbreeds" with him.[5]

The oversized shadow of Nero Wolfe fell across the detective genre at exactly the time when Raymond Chandler was trying to break out of pulp magazines like *Black Mask* into writing an almost entirely new kind of detective fiction. The English-educated Chandler was sensitive, even snobbish, about his work, refusing to allow it to be published alongside those who, "whatever their smoothness and accomplishment in their field," simply "do not matter as writers, except in a purely commercial way. To put oneself in that company is to classify oneself, and the dough aint [*sic*] enough." His goal was not to become a serious novelist, but to take a debased vernacular form and make it into something new, rich, and distinctive.

Although Chandler needed to see his work promoted and widely read, he told his publisher, "do as little as possible to lump me in the public mind with the smooth and shallow operators like . . . Stout." On another occasion, he told a fan who had called him a "great gentleman," that she was mistaken: "If you count me one a par with Erle Stanley Gardner and Rex Stout, I am very sorry. I count myself far above them. This will prove to you that I am not a gentleman at all."[6]

Chandler made his goals as a writer clear in a classic essay called "The Simple Art of Murder" (1950), which is where his celebrated definition of the ideal private eye hero ("a man who is not himself mean, who is neither tarnished nor afraid") comes from. In Chandler's view, murder had to be taken out of the country house, away from the titled detective (or his risible American counterfeit) and put back in sleazy tenements, small sad offices and on the mean streets. Not everyone liked this, of course, but Chandler casually dismissed the style's critics as "flustered old ladies — of both sexes (or no sex)," who object to unsentimental realism because they "like their murders scented with magnolia blossoms" (or orchids).

In the context of Chandler's ambitions, the role played by orchids in *The Big Sleep* makes sense. Chandler's description of the greenhouse is powerfully atmospheric, full of rich metaphors and potent imagery, but there's almost nothing about the orchids themselves; no details of how many species there are, nor of how they're grown or why. Marlowe doesn't care and neither do we. By contrast, Wolfe's orchid house is described in considerable detail across many books and stories: rooms were set aside for *Cattleyas*, and hybrids, while others held

Odontoglossums and *Oncidiums*. Ten thousand flowers bloomed in an elaborate setting of angle-iron staging painted in gleaming silver and several species are named and described accurately.[7] However, unless you're passionate about the orchids themselves (as Stout doubtless was), these descriptions add nothing to the books, while Chandler uses a single stripped-down scene of orchids to engage the reader's emotions immediately.

General Sternwood's orchid greenhouse is a tour de force; the writing is as tight and memorable as any in Chandler's works. The scene was surely intended to take orchids back from Rex Stout, and offer a lean alternative for Nero Wolfe, whose bloated, saggy features perfectly matched his book's bloated, saggy plots. It also has to be acknowledged that these two authors embody competing visions of masculinity; Wolfe seems distinctly effeminate in his tastes and behavior, and *The Big Sleep* is gratuitously and casually homophobic. When Chandler famously described his ideal detective in the conclusion to "The Simple Art of Murder," he specified that he must be both honorable and heterosexual ("he might seduce a duchess and I am quite sure he would not spoil a virgin") and he must also be "a relatively poor man, or he would not be a detective at all," as well as a common man with a "disgust for sham." In other words, he must be the exact antithesis of everything Nero Wolfe embodied. Using orchids would have brought Wolfe to mind for many of Chandler's readers, but by stripping them of all the mannered and eccentric padding, he showed that less is more; he was going to rescue murder from the orchid fancier, who pursued it only to finance his effete addiction, and give it "back to the kind of people that commit it for reasons."

Thanks in large measure to Stout and Chandler, orchids became a recurring feature of the kind of hard-boiled detective story that Dashiell Hammett pioneered and Chandler perfected. The flowers were often used to evoke a vulnerable, rather artificial world of luxury and decadent ease. Most examples are as forgettable as Stout's stories, but one of the more interesting is James Hadley Chase's notoriously violent and massively successful novel *No Orchids for Miss Blandish*, which appeared in the same year as *The Big Sleep*. Orchids are never mentioned once in the novel, yet they had become such familiar props that Chase's title was enough to successfully conjure the expensive, fragile quality of the heiress, Miss Blandish, who not only seems to lack a first-name but (like one of Sternwood's daughters) has "never had any sense of values," merely "enjoyed a good time all the time," until she is kidnapped and raped by a psychotic gangster, aided and encouraged by his mother. Once rescued Miss Blandish kills herself and, in one of the most unpleasant twists in a very unpleasant novel, it's implied that she is unable to live without the man who has abused her, as if she'd become addicted to abuse. In the 1948 film version, the director, St. John L. Clowes, tried to address Chase's apparent oversight of omitting to include any actual orchids, so in the film's final scene we glimpse the now dead Miss Blandish's orchid corsage being trampled by an indifferent public. The film was so violent (by the standards of the day) that Britain's future prime minister, Harold Wilson, who was the government minister responsible for films, announced publicly that he delighted to see there had been "no Oscars for Miss Blandish." The novel was also condemned by George Orwell, who saw it as epitomizing the entertainment of a fascist age; brutal, power-hungry, and sadistic (but reading his essay, one has the impression that he was at least equally offended by its Americanness and the idea that it might teach English readers to speak in vulgar American slang).

The absent orchids of Chase's novel stood for the wealth, luxury, and innocence that Miss Blandish loses, but they were also used to signify the ultimate bouquet or accolade. The year after Chandler's and Chase's novels appeared, Edward G. Robinson starred in a movie called *Brother Orchid* (directed by Lloyd Bacon, 1940), in which he plays Little Johnny, a double-crossed gangster forced to hide out in a monastery where the monks grow flowers to raise money for the poor. As an alias, he takes the name in religion of Brother Orchid, because orchids signify the "class" he's been trying to buy himself with his ill-gotten gains. The original trailer for the movie used the phrase "Orchids to . . ." several times, to signal the bouquets that would be heaped on Robinson and his costars, as well as offering the audience "orchids to the swell time you're going to have."[8] This phrase was widely used at the time: there was a 1935 movie called *Orchids to You*, and the magazine *Theatre Men's Guide* used the phrase to praise the original version of *A Star is Born* (1937). The phrase was even in *The Big Sleep* (in the book's only other orchid reference): when Marlowe is knocked out by some gangsters he awakens to find he has been tied up and is being watched over by the head gangster's wife; he complains about the pain and she replies, "What did you expect, Mr. Marlowe—orchids?" Marlowe responds, "Just a plain pine box."[9] Orchids meant applause, so Chase's title suggests there would be no plaudits for Miss Blandish; her vacuous life of pleasure would bring her only the most horrible misery. Chase wrote a sequel to Miss Blandish called *Flesh of the Orchid* (1948), which reintroduces the fin-de-siècle's clichéd predatory, sexy woman by inventing a daughter for Miss Blandish (quite when and how she was born is one of several holes in the plot). The daughter has a split personality because she's a mixture of her psychotic father and good-time mother, and the weak book's title is justified by having the daughter, Carol Blandish, leave a blood red orchid at the scene of each of her crimes.

Frail Orchids

For the hard-boiled writers, the symbolism of orchids was as showy and obvious as the species they chose to write about; the big, flamboyant tropical orchids are the ones typically splashed across the covers of the classic detective writers. Yet while the prized tropical species might seem rampant, even dangerous, in their native habitats, they become delicate blossoms once they've been transplanted to a temperate greenhouse, where they need an expert's tender care if they are to survive (the very difficulty of keeping them alive added to their nineteenth-century allure). And many other orchids are inconspicuous; the orchids of temperate regions are typically tiny and their flowers are often too small even to be noticed by a casual passer-by. In the twentieth century, the various frailties of orchids would be used to convey a richer variety of ways in which men might be men.

Britain's native orchids are hard to see, they seem almost shy, and the apparently secretive qualities of these unassuming little flowers caught the imagination of the Kentish novelist and naturalist Jocelyn Brooke. It was not tropical orchids he was interested in, but Britain's indigenous species, which he collected in the Kent and Sussex countryside, just as Darwin had done a century earlier. In *The Wild Orchids of Britain* (1950), Brooke noted that "like certain other blameless forms of vegetable life," orchids had been "unfortunate in their associations," whether it be with wealth, ostentatious luxury, or "the recondite tastes of a des Esseintes or the slightly vulgar exoticism of a Wilde." Brooke clearly hoped to rescue these flowers from such unsavory company and associations. The book's delicate watercolors seem to have been drawn in deliberate contrast to the rather overpowering illustrations in the great Victorian orchid monographs, as if to assist in his mission of turning his readers' minds away from "perilous expeditions through miasmal jungles," or "hairbreadth escapes and triumphant discoveries in the manner of the *Boys' Own Paper*." Although many British orchids were so rare that they presented a real challenge to even the most expert collector, he insisted that "there is nothing very decadent or *fin-de-siècle* about any of them."[10] Britain's orchids sounded almost boring in Brooke's description, yet in his semiautobiographical novels he used the flowers to illustrate a fascinating and very different manliness to that personified by the heroic, tropical collector.

Brooke was a shy, awkward boy who hated school and preferred to write poetry and roam the countryside looking at wildlife. Feeling himself unsuited to taking over the family business, he joined the Royal Army Medical Corps (RAMC) when WWII started and became what was known as a "pox wallah," a specialist in the treatment of venereal diseases. Something about the camaraderie and order of army life must have appealed to him, because after the war, still unable to find a job he felt suited to, he rejoined the army, remaining in the RAMC until the success of his first, largely autobiographical novel, *The Military Orchid* (1948), allowed him to buy his way out. (The novel became part of *The Orchid Trilogy*, with the addition of *A Mine of Serpents*, 1949, and *The Goose Cathedral*, 1950.[11])

Brooke's novels tell the simple tale of a life very like his own; his main character has an idyllic childhood, engages in numerous (usually unsuccessful) hunts for British orchids, and spends time in the army. A particular focus of his interest was *Orchis militaris* (the "Military Orchid" of the title, so called because its flower looks a little like a helmet), and careful reading of Brooke's subtle, delicate prose reveals that his narrator's interest in the plants was more than botanical. For example, when he wrote about the land near Dibgate, Kent (where the British army still maintains a training center), he described it as "a frontier-land, inhabited only by soldiers":

> Sometimes I encountered these frontier-tribes: parties of troops in training, out on a cross-country run. In shorts and singlets, they plodded heavily across the rough fields, their naked limbs stained purple by the shrewd east wind, their red faces set in an expression of dull, stoic endurance. They seemed some curious variation of the Human species—*Homo sapiens* var. *militaris*—indigenous, like the Military Orchid, only in a few isolated, calcareous [chalky] districts off the beaten track . . . Remote and alien, they passed me by without greeting, no flicker of human emotion betraying itself in their coarse, meaty faces. Watching them, I felt a stranger from another world: barred implacably from any friendly contact with these denizens of an alien country. (248–49)

Why does the orchid collector feel so isolated from his species, particularly its uniformed members? The novel's writer and readers gradually become kin to the private eyes, searching for elusive clues to the narrator's own nature, trying to make sense of the hints scattered through the book. While Neville Chamberlain was busy appeasing Hitler at Munich, the narrator "embarked upon a full-length monograph on the British Orchidaceae" (as Brooke did in real life); he regards this quixotic undertaking as "fiddling while Rome burnt." With the world on the brink of war, the orchid monograph provided a pretext to ignore politics and continue the search for the elusive Military Orchid. The narrator rents a room in a boarding house at Dover, which reminded him of a line of W. H. Auden, "The cry of the gulls at dawn is sad like work." He noted that "as it happened, Messrs Auden and Isherwood had recently stayed there too: '*they* was a pair of scamps, if you like,' the landlady alleged." (Sadly, no details of how the pair justified this description are given.) Meanwhile he'd heard rumors that the Military Orchid had been seen nearby, but it "remained, as it probably always would remain, unfound" (257).

There is something haunting about this long, low-key story of a shy man on a seemingly hopeless quest for secretive orchids that are like soldiers and soldiers who are like orchids. In a later scene, the writer and another botanist go looking for a rare orchid near ancient long barrows, Neolithic burial mounds, "tall beeches sprang from the tombs, their roots imprisoning more firmly, year after year, the warrior-bones below." They find yet more soldiers on maneuver, but "the living soldiers marched away; only the dead remained" and they found their orchid, which Brooke identifies as *Epipactis helleborine* (Broad-leaved Helleborine). "Its flowers were not open: they never would open, for the Helleborine had long ago solved the sexual problem for itself by becoming self-fertile." The metaphorical meanings of orchids take a rich new turn:

> Not only did the flowers not open, but the thin, papery lip withered in the bud, the rostellum was evanescent. The Helleborine seemed to have given up the unequal struggle: it was a case of floral introversion, an evolutionary retrogression to the auto-erotic phase. (258)

The combination of "introversion" and "auto-erotic" provide clues to one of the hidden meanings in Brooke's story. Conventional psychology in the 1940s was wedded to a version of Freud's theory that sexuality progressed through distinct phases, including an auto-erotic one, eventually evolving via a homosexual phase into adult heterosexuality. "Introvert" faintly echoes "invert," a term for homosexual that was still in regular use at this time, when many psychologists would have explained homosexuality as the result of a person having become stuck in their sexual development (perhaps through shyness), or—as Brooke's passage on the Helleborine implies—regressing to an earlier phase of development. From scattered clues like this, what gradually emerges is the story of Brooke coming to terms with his homosexuality at a time when it was still illegal. He has no desire to emulate Wilde or his flamboyant heroes, and no desire to shock. His quest for the Military Orchid is a search for himself and for love, ideally in the arms of a fellow soldier.

Brooke's protagonist finds one of the rarest of all orchids while in the army, in Italy. As he's searching for flowers he almost trips over a company of Italian soldiers: "their grey-green uniforms merged in the undergrowth, they lay immobile as lizards beneath the chequered shade." They are so well camouflaged that they become part of the landscape:

> They seemed the indigenous fauna of this Tuscan countryside: shy, watchful creatures, not dangerous, but none the less faintly inimical; goat-footed, possibly, and covered, below the navel, with thick, matted hair . . . One of them grinned at me as I passed: a sudden flash of white teeth in a patch of brown shade. (269)

And close by the soldiers, he finds "the Lizard Orchid: that legendary flower, most celebrated of English 'rarities,' which I had sought for, year after year, on the Kentish chalk-lands—but without success." The description of the indigenous soldiers, "immobile as lizards" links them to the orchids:

> To pick a Lizard Orchid—the action had about it something unholy, something rather blasphemous . . . I picked several more; examining with a wondering delight the long, slender lips, two inches long, cleft at the tip like serpents' tongues, unfurling themselves in delicate spirals from the opening buds. (270)

Again "something rather blasphemous" reminds us of the pagan, goat-footed soldiers. Brooke is understandably discreet about the sources of that Tuscan afternoon's happiness, but the reader is left hoping that the orchid's weren't the only "slender lips" he found.

Twentieth-century orchids could be used to suggest many kinds of masculinity, in many kinds of fiction. Flamboyant tropical species featured regularly alongside the tough heroes of the classic hard-boiled detective story, where they tended to retain their nineteenth-century association with predatory, overperfumed bad girls. By contrast, Britain's elusive native orchids were used in Brooke's fiction to hint at a love that still could not speak its name at midcentury. The increasing diversity of orchid associations meant they could be used to convey a very different kind of detective, and a different kind of manliness, in Norman Jewison's movie *In the Heat of the Night* (1967).

The film is set during a baking hot summer in Sparta, Mississippi. Sidney Poitier plays Virgil Tibbs, a Philadelphia homicide detective who's passing through on his way home and has to change trains in the middle of the night in the middle of nowhere. One of the local white cops discovers a murdered man on Main Street and when he finds a strange black man — suspiciously well-dressed and with money in his pockets — at the railroad station, leaps to conclusions and arrests him. Once Tibbs' identity is established, he helps the white police chief, Gillespie (played by Rod Steiger), to solve the crime. In a slightly implausible celebration of the liberal hopes of the late sixties, Gillespie's redneck suspicion is initially aroused by Tibbs' polished, big-city manners and clothes, but they gradually learn to respect and eventually even to like each other.

As Tibbs investigates the murdered man's car, he finds a piece of *Osmunda*, a kind of fern root, on the brake pedal, which he recognizes because it's used for potting orchids.[12] He leaps to his own racially motivated conclusions and identifies Eric Endicott (Larry Gates), a passionate orchid grower, as the prime suspect. Endicott owns cotton plantations, and is firmly racist in his views. By contrast, the murdered man, Mr. Colbert, was a liberal Northern industrialist who was planning to build a nonsegregated factory in Sparta. So Tibbs assumes that Endicott had a motive for murder, to stop a development that would forever free his workforce from picking cotton.

Gillespie and Tibbs drive up to Endicott's mansion, passing through fields where black people are picking the cotton exactly as they did a century earlier. "None of that for you, huh, Virgil?" jokes Gillespie (Poitier's speechless reaction shot is superb). When the two police officers arrive at Endicott's mansion they are shown into his orchid greenhouse and a black butler brings them lemonade. Gillespie is street-smart and tough but insecure about his abilities, whereas Tibbs' self-confident masculinity is more cultured and educated, as he demonstrates to Endicott by displaying his detailed knowledge of orchids. The ensuing discussion of the white man's flowers is fascinating:

Endicott: Have you a favorite, Mr. Tibbs?

Tibbs: Well, I'm partial to any of the epiphytics.

E: Why, isn't that remarkable! That of all the orchids in this place, you should prefer the epiphytics. I wonder if you know why.

T: Maybe it would be helpful if you'd tell me.

E: Because, like the Negro, they need care and feedin' and cultivatin'—and that takes time. That's somethin' you can't make some people understand. That's somethin' Mr. Colbert didn't realize.

Tibbs (young, intelligent, handsome, and towering over the elderly Endicott) listens calmly to this frail, patronizing relic of the old South, as he insists that his "Negroes" are like orchids, transplanted from the tropics to a hostile environment where they cannot survive without the expert care of a rich white man. Tibbs responds by demonstrating his own, orchid-based expertise: he picks a piece of fern root from one of the orchid baskets and asks: "Is this what the epiphytics root in?"

E: My point! They thrive on it. Take it away from them, they do poorly.

T: What do you call this material?

E: That's *Osmunda.* Fern root.

Gillespie, who has been sitting, apparently bored, throughout this discussion, perks up dramatically at the mention of *Osmunda* and suggests it's time to leave. Endicott becomes suspicious and Tibbs confirms he's a suspect, asking him, "Was Mr. Colbert ever in this greenhouse? Say, last night, about midnight?" whereupon Endicott slaps his face—and Tibbs slaps Endicott right back. This moment, probably the first time an African-American had struck a white man in a mainstream US movie, has been called "the slap heard around the world." In the largely segregated US cinemas in which the movie opened, white audiences sat silent and stunned while black ones cheered Poitier's cool, insouciant assertiveness.[13] In the movie, Endicott turns to Gillespie and asks, "You saw it?" to which the police chief replies, "Oh, I saw it." "Well," asks Endicott, "what are you gonna do about it?" and Gillespie, as confused as everyone else (apart from Tibbs) says, "I don't know." (As the town's mayor later points out, their last police chief would simply have shot Tibbs and claimed self-defense.) The policemen leave the greenhouse with Endicott in tears and the black butler visibly shocked at Tibbs' lack of deference. It later turns out that Tibbs is wrong; Endicott had nothing to do with the murder, but this mistake is a turning point in his relationship with Gillespie. However, it's the slap amid the orchids that stayed with the viewers. The self-evidently ludicrous image of Sidney Poitier as an orchid — frail, vulnerable, and in need of white care — is one of many things that gives the orchid house scene from *In the Heat of the Night* its emotional impact. The orchids add another dimension to the film: Tibbs is no less of a knight errant than Chandler's Philip Marlowe, every bit as willing to risk his life in the pursuit of justice, but he's a man who relies on his brains rather than his fists, whereas Gillespie would be much more at home tackling the mean streets in the old-fashioned way. And Tibbs is a man who knows his orchids; they dramatize the contrast between his cultured expertise and Gillespie's redneck ignorance.

Loud, flamboyant men can be sent off to the jungle to collect loud, flamboyant orchids, but the delicacy and inaccessibility of orchids have also been used to suggest the personalities of shy men, who might not be quite capable of surviving or thriving in the everyday world — much less in the fetid jungles. And the orchid's enigmatic sexuality has been used to suggest other dimensions to these quiet, secretive men. For example, the Australian film *Man of Flowers* (directed by Paul Cox, 1983), opens — with no preliminaries or explanations — with an extraordinary scene in which a startlingly beautiful young woman does a coyly sexy striptease to an aria from Donizetti's opera *Lucia di Lammermoor*. She is watched by a balding, middle-aged man, who we later learn is Charles Bremer, an eccentric and wealthy lover of art and flowers. As soon as the performance is over, Bremer flees the room in order to hammer out some rather deafening avant-garde improvisations on the local church's organ. As the audience will discover, he is almost incapable of expressing his feelings, most notably his sexual ones, in any other way. It's a blackly comic opening to a bizarre film.

The audience soon discovers that the stripping girl is Lisa (played by Alyson Best), a life-model at the evening classes where Bremer is studying art. She does some additional undressing to music at Bremer's home because she needs the money for her boyfriend, David (Chris Haywood), an unsuccessful painter with a major drug habit. When David urgently needs $1,500 to pay his dealer, he bullies Lisa into approaching Bremer for more money. After one of the life classes where she's been modeling, she tells him, "I'm in trouble—or rather, David's in trouble—and I need a lot of money. I'll do anything." She offers to bring her lesbian friend Jane around for the weekend, promising again that they will "do anything" for the money.

Bremer notices for the first-time that she has a black eye and is obviously scared of David. He tells her: "And next time he'll want more money, won't he? [*pause*] My poor little flower; how can you go on like this? Yes, yes. Come over the weekend. Bring your friend. We'll make it two thousand." As he starts to walk away, he adds, "Spend the change on something useful," pauses for a moment and suggests, "Buy some orchids. Every house looks better with orchids." No other flower would make this moment funnier; it perfectly captures Bremer's peculiar innocence that makes this otherwise rather creepy character strangely endearing. It takes a degree of other-worldly detachment to view these proverbially expensive, exotic, hothouse flowers as "something useful." The next scene shows him arming a lethal trap with which he will solve Lisa's problems; he calmly murders David, encases him in bronze, and presents him to the city of Melbourne as a public statue. Sex, death, and orchids are once again tightly linked.

Orchids also played a small supporting role in *Secretary* (directed by Steven Shainberg, 2002), which centers on the relationship between Lee Holloway (played by Maggie Gyllenhaal) and her boss, Mr. Grey (James Spader). Each is shy, lonely, and incapable of expressing conventional sexual feelings. They eventually find happiness in a consensual sadomasochistic relationship (the negotiation of which is comic, touching, and ultimately profoundly liberating for both of them). Mr. Grey's life is as colorless as his name; his fussy, precise manners are echoed by his overly neat and ordered office. The one splash of color is provided by the orchids that he tends with an obsessive routine. The feelings he finds so difficult to admit or express are sublimated into the care of these constrained, overcultivated flowers.

The flowers Mr. Grey grows are *Phalaenopsis* (commonly known as Moth Orchids), the most widely cultivated group of orchids today. Despite their popularity, some people loathe them, perhaps because their bare branches— surmounted by a few, isolated blooms—suggest a pinched artificiality. *Phalae-*

nopsis are native to Southeast Asia and Japan, Taiwan, and Thailand are the world centers of modern, high-tech orchid growing (and of the continued destruction of wild habitats by orchid pirates). Reactions to these orchids (positive and negative) are perhaps colored by their being considered "oriental" — just as some people like the precise delicacy of bonsai trees while others think they look tortured, more reminiscent of foot-binding than of natural beauty. The slightly unnatural quality of *Phalaenopsis* is what makes them such apt flowers for the uptight Grey to obsess over before his secretary helps him find a more pleasurable outlet for his passions.

Orchids seem to have changed gender regularly in our cultural imagination and their sexual significances have been even more fluid. In the eighteenth century pseudo-classical myth that bears his name, Orchis was imagined not only to be male, but also as personifying the most violent and unpleasant aspects of male sexuality. In the following century, that image faded gradually as orchids began to symbolize new stereotypes for women, as vamps and seducers (perhaps reflecting unease over women's gradually growing demands for equality). And of course they were also used as decadent symbols of "the recondite tastes of a des Esseintes or the slightly vulgar exoticism of a Wilde." It was partly in reaction to the effeminate orchid that the imperial orchid-hunting genre emerged to reassert traditional masculinity by using the flowers as trophies and emblems of manly courage and hunting prowess. As the twentieth century unfolded, orchids became ever-more sexually ambiguous, since they could symbolize either women becoming more masculine (more like Orchis) or men becoming more feminine (like the shy and gentle Brooke). Yet these contradictory uses, rather than diluting the potency of orchids as sexual symbols, seem to have enriched it; what united these seemingly contradictory associations was a kind of sexual dissidence — orchids could be used to symbolize anyone who refused to fit conventional sexual roles (hence their association with Mr. Grey in *Secretary*). Most surprisingly of all, it was thanks in part to the diverse ways in which the sexuality of plants had been reimagined in fiction, that twentieth-century science would discover that the realities of the orchids' sex lives were much kinkier than anyone had ever suspected.

10

Deceptive Orchids

In addition to novels and poetry, Jocelyn Brooke wrote about the natural history of orchids and understood their biology well, as his discussion of the Helleborine's regression to self-fertilization illustrates (see previous chapter). This was something Darwin had also explored; why, when so much biological energy and evolutionary time has gone into adaptations to ensure cross-fertilization, do some orchids give up and revert to self-fertilization? For example, the Bee orchis (*Ophrys apifera*), which both Darwin and Brooke saw growing on the Kentish hills, is seldom visited by the bees it resembles, despite being clearly adapted for insect pollination (it even retains the sticky glands to glue the pollinia to the bee). Darwin speculated that in such cases, the orchid's pollinator had failed it; perhaps it became extinct, the orchid migrated to regions where the pollinator was rare, or rival flowers had competed more successfully for the insect's attentions. In these cases, as in Brooke's Helleborine, natural selection would modify the flower to ensure that self-fertilization might occur: "For it would manifestly be more advantageous," Darwin noted, for "a plant to produce self-fertilised seeds rather than none at all or extremely few seeds."[1] In other words, self-fertilization is better than no fertilization at all, as Brooke's writing seems to hint.

However, there was a great deal about orchid fertilization that was still mysterious in the twentieth century, despite the decades of research since Darwin's day. Among the most interesting was that of mimicry; why did the Bee orchis look like a bee in the first place?

In 1829, an Oxford don called Gerard Smith published a slender *Catalogue of Rare or Remarkable Phaenogamous Plants, Collected in South Kent* ("phaenogamous" means "visible marriages," an old botanical term for flowering plants as opposed to those Linnaeus called "cryptogamous," i.e., "hidden marriages," meaning the flowerless plants, like mosses). Smith's book was one of many consulted by Darwin when he was doing his orchid research, and (as we saw in chapter 5) he was puzzled by the following statement:

> The Bee-Ophrys has, indeed, the appearance of that insect, engaged in pilfering a flower; Mr. Price has frequently witnessed attacks made upon the plants by a Bee, similar to those of the troublesome *Apis muscorum*; and I have myself seen a young entomologist, approaching stealthily, with out-stretched hand, the successful deceiver, whose mimic beauty became, alas! its own ruin.[2]

As Smith noted, "an extraordinary share of wonder" attached to these strange orchids, "the 'sports of nature' as they are playfully called, as though creation could trifle." As noted above, Darwin quoted part of Smith's *Catalogue* in a footnote to his book on orchids and commented on the bee apparently attacking the flower, "What this sentence means I cannot conjecture."[3]

Darwin's puzzlement is hardly surprising; he never saw any insect visit the Bee orchis, which is invariably self-fertilized in Kent, but in any case why would a bee "attack" an orchid? And how could natural selection shape an orchid so that it resembled an insect so successfully that an entomologist would try to catch one? Linnaeus himself wrote about another Ophrys, *O. insectifera*, whose "flowers bear such a resemblance to flies that an uneducated person who sees them might well believe that two or three flies were sitting on the stalk."[4] Darwin was also baffled by a second aspect of orchid biology; many species did not appear to produce nectar with which to reward their pollinators. This fact had also puzzled the eighteenth-century German botanist Christian Konrad Sprengel, who wrote one of the first significant works on insect pollinators, *Das entdeckte Geheimnis der Natur im Bau und in der Befruchtung der Blumen* (The Secret of

Nature in the Form and Fertilization of Flowers Discovered, 1793). Darwin read this carefully and found many of Sprengel's insights valuable, but was unable to agree with the German's conclusion about the flowers he called *Scheinsaftblumen* (sham-nectar-producers), which was that they relied on what Darwin called "an organized system of deception." Why, the Englishman wondered, would insects continue to visit and pollinate if they received no reward? As Darwin wrote, anyone who believed the insects could continue to be fooled, generation after generation, "must rank very low" the instincts and intelligence of many polli-nating insects.[5]

Despite Darwin's celebration of plants as "wonderfully crafty," he never fully realized just how cunning they really were; it would not be until the early twen-tieth century that a French judge, an English colonel, and an Australian school-teacher finally solved the twin puzzles of insect mimicry and the success of the "sham-nectar-producers," partly because they realized these were one puzzle, and partly because they realized orchids were more seductive and devious (in fact, much more like their fictional counterparts) than even Darwin had ever suspected.

The story begins with Swiss botanist Henry Correvon, a pioneering conser-vationist and president of the Swiss League for the Protection of Nature (which helped save the country's national flower, Edelweiss, from extinction at the hands of overenthusiastic tourists).[6] He was also one of many who were inspired to investigate their local orchids after reading Darwin's work on orchids (which was so successful that it soon appeared in translation). However, some of the *Ophrys* species he was interested in did not grow in Switzerland, so he wrote to one of his many correspondents, Maurice-Alexandre Pouyanne, a French colo-nial judge in Algiers, presiding over Sidi-Bel-Abbès (home town of the French Foreign Legion). Pouyanne was one of a number of French botanists who took advantage of France's African empire to study tropical plants; at Correvon's sug-gestion, he began to spend his (apparently rather ample) spare time studying the pollination of North African orchids by insects, particularly those like *Ophrys* that mimicked insects and whose fertilization was still not understood. Pouy-anne's conclusions were so surprising, even to him, that he spent twenty years on painstaking research before he was confident enough to publish.[7]

Pouyanne's work focused on a species of orchid called *Ophrys speculum*, whose name means "mirror orchid" because its labellum has a metallic blue, glossy lip that seems to reflect the Mediterranean sky under which it grows. Rather like the bee orchid, *Ophrys apifera*, that had puzzled Darwin, *O. speculum* has a lip that is fringed with long, reddish-brown strands that are the modified margins of the petals but look like the hairs found on the abdomens of many insects. These orchids were visited regularly by burrowing wasps, but they made no attempt to feed on the orchids (unsurprisingly, as they are carnivorous). To investigate their visits, Pouyanne described how an experimenter needed to sit in the sun, "a small bouquet of *Ophrys speculum* in your hands," as if waiting to court the wasps. Once the insects appear they will struggle to gain possession of a bloom, but once one has landed on the orchid:

> its abdomen dives at the bottom in the long red hairs that look like the labellum's bearded crown. The abdomen tip becomes agitated against these hairs with messy, almost convulsive movements [*les mouvements désordonnés, presque convulsifs*], and the insect wiggles around.[8]

And after the peculiar movements, the pollinia of the orchid end up glued to the insect, from where it would, of course, be transferred to the next orchid that the wasp entered in this peculiar way.

As Pouyanne investigated this strange behavior, he noticed some interesting facts: all the wasps were males and those messy, "convulsive" movements only occurred in the early weeks after the first wasps hatched—when there were no females around. In his efforts to try to identify the source of the attraction, Pouyanne tried all kinds of things. He found that if he cut off the orchid's labellum (the part that most resembled an insect), but left all the other organs intact, the insects ignored them completely. That seemed to confirm the importance of the visual similarity; the shiny color of the flowers resembled the iridescence of the female wasps—not very closely, admittedly, but (as Pouyanne argued) insects are very shortsighted. His idea was confirmed when flowers on the ground attracted insects; even upside-down flowers worked, although less effectively. However, while his hand-held bouquet attracted males, if the same bouquet were forced on the females they proved indifferent to the orchid's charms and if the experimenter pressed his suit too insistently, the females flew off as if repelled. If the orchids' appearance attracted one sex, why not the other? Pouyanne began to suspect that the orchid's scent, almost imperceptible to humans,

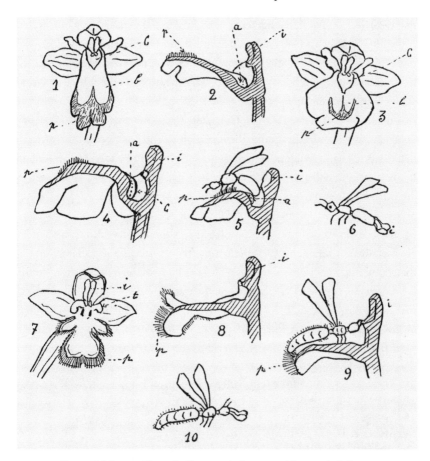

Figure 40. Maurice-Alexandre Pouyanne's drawings of the wasp's abdomen
displays "almost convulsive movements" as it fertilizes an *Ophrys* orchid.
(From Pouyanne and Correvon, "Un Curieux Cas de Mimétisme chez
les Ophrydées," *Journal de la Société Nationale d'Horticulture de France*, 1916)

was the key. To test his idea, he hid flowers under sheets of newspaper, and—as
he had hoped—found that the males went looking for them anyway.[9]

Pouyanne's conclusion was astonishing. When he described his wasps, he
put it very plainly: *"ses mouvements, son attitude paraissent tout à fait semblables à
ceux des insectes qui pratiquent des tentatives de copulation"* (his movements, his
attitude seemed quite similar to those of insects attempting copulation).[10] The
insects were attempting to mate with the orchids, which had evolved to exploit

that brief delay between the hatching of the males and females. Pouyanne's first paper appeared (with Correvon's help) in February 1916 when France, like the rest of the world, had slightly more pressing matters than orchid pollination to worry about; it would be several more years before the significance of Pouyanne's work was fully recognized.

Just a few years later Edith Coleman, an Australian amateur naturalist, independently discovered a similar phenomenon; as she described it, "an interesting, but perplexing problem" had presented itself after her daughter "described to me certain remarkable actions on the part of a wasp," that had landed on an orchid, but "entered the flowers *backwards*." She went to see for herself and not only confirmed her daughter's observation but added several others that "have so far puzzled several leading entomologists." Although the flowers seemed almost scentless to humans, something attracted the wasps and Coleman noted that the orchid's lip "somewhat resembles the body of the insect visitor," to the extent that its curvature "exactly meets the needs of the wasp," a resemblance that led her to "hazard a theory." She noted that the orchids and wasps seemed locked in an embrace; it took "an apparent effort" for the wasp to free itself, "which shook the flower." But the fact that the wasps were not "deterred" by their apparent discomfort, "would suggest that it has in some way received payment for its former visits."[11] It seemed that the Australian wasps were also trying to mate with the orchids.

Within a year of her first paper, Coleman had discovered the earlier work of Pouyanne and of a British naturalist, Colonel Masters John Godfery, who had confirmed the Frenchman's findings (according to his obituary, his key research "was carried out by watching cut flowers in vases on hotel verandas in the Mediterranean region").[12] Coleman was then able to publish a more detailed account of her Australian orchids, in which she concluded that the wasps were "seeking neither nectar nor edible tissue, they are answering to an irresistible sex-instinct." The wasps had been fooled by a "cunning mimicry":

> The strange labellum, modified out of all proportion to the thin, almost thread-like petals and sepals, with its double row of dark glistening glands that gleam in the hot sunshine loved by the wasp, is perhaps sufficient to justify the theory of an attraction based on the resemblance of the flower to a female wasp. Even to our eyes, the likeness is apparent. To the inferior eyesight of the insect, the resemblance may be still more convincing.

Figure 41. Edith Coleman was among the first to use photography
as part of her botanical examinations, including stereographic images.
(*Cryptostylis erecta*, photograph by Edith Coleman and Ethel Eaves, 1931)
Credit: State Library of Victoria.

However, Coleman was convinced that the visual resemblance alone was in-
sufficient to explain the attraction. She noted that when flowers "were placed on
a shelf under an open window" they were "almost instantly" visited by wasps;
"the fact that a single flower will lure an insect from a distance, would suggest
that, even though the perfume is so subtle as to escape our notice, it is very
readily conveyed to a wasp."[13] She was the first to conclusively prove her theory
by identifying wasp semen in the flower, and she also pioneered the use of photography to investigate and illustrate the phenomenon.[14]

Pouyanne, Godfery, and Coleman seem to have arrived at the same conclusion independently: through a combination of scent and visual mimicry, orchids
are able to fool male wasps into thinking their flowers are female wasps. Natural
selection had modified the flowers' appearance and scent just enough to lead the
male wasps astray, obtaining their services as pollinators while giving nothing in
return. (In the 1950s, biologists finally identified the chemicals in the orchid's

scent that completed the mimicry; the orchids may not have had much scent to human noses, but they produce chemicals that mimic the pheromones, the sex hormones, that female wasps emit.[15]) Coleman described the action of orchids as follows: "With a Machiavellian cunning almost beyond belief, the orchid lures the artless insect to its service" and commented that "to follow closely the act of pollination increases one's belief in the sagacity of plants."[16] As the distinguished historian of botany, the late William Stearn, commented, the crimes attributed to orchids by J. E. Taylor (chapter 6) are nothing compared to the sexual deception practiced on the hapless wasps.[17]

In the 1930s, the great American orchid biologist Oakes Ames was the first to write the history of the discovery of this bizarre form of pollination, which has become known as pseudocopulation. He acknowledged that (at that time) almost nothing was known about how it evolved, but concluded:

> It may be that those who would reject the evolutionary approach to an understanding of life and who prefer to regard the world as the product of Special Creation will lean a little more lightly on human weakness when they discover moral turpitude among the insects. . . .[18]

—to say nothing of the morals of the orchids. Ames even wondered whether the discovery of pseudocopulation would encourage entomologists to "become Freudian in their outlook, when discussing the sexual vagaries revealed" by scientists, perhaps introducing terms such as *Ophyrdean* complex" into their analyses. "Perhaps," Ames concluded, "even the poet will have to reconsider whether 'only man is vile.'"[19]

Darwin himself failed to spot this vital evolutionary relationship, perhaps because—like most men of his day—he was unable to imagine women being either as intelligent or as sexual as men. Darwin assumed that the intense evolutionary competition between males made them more intelligent than females, so he sent his sons to university, but not his daughters. And when he was trying to decide whether or not to marry, he weighed the attractions of a "nice soft wife" against the "Conversation of clever men at clubs," and although marriage won out, Darwin's notes make it clear that it never really occurred to him that his potential mate, Emma, might have her own views on the matter.[20] Given how common these preconceptions were in Darwin's day, it's not surprising that neither he nor his contemporaries realized just how wily and seductive the orchids were. It seems that both orchids and women would have to be thought of in very different ways before the truth could be grasped.

By the early twentieth century the sex lives of orchids, whether real or imaginary, revealed far more complexity than even Darwin had ever imagined. Whatever motivated Pouyanne, Godfery, and Coleman to investigate this aspect of orchids, it's striking that they were able to see what Darwin himself had missed. A full understanding of the orchid's pollination strategies seems to have required two things: first, Darwin's work influenced others into reimagining plants as imbued with cunning, able to deceive (and even to devour) insects. But was it more than coincidence that the orchids' ability to trick male wasps into pollinating them wasn't understood until after various writers had reinvented the orchid not merely as feminine but also as a wily seductress? That second step depended on women challenging age-old stereotypes, which suggests that real and fictional orchids — the worlds of the pulp writers and the serious scientists — were not as far apart as they might appear; the stories of real and fictional orchids turn out to be part of a single story.[21]

Orchids in Orbit

The first hints of a fertile cross-pollination between science and science fiction are, unsurprisingly, to be found in the work of H. G. Wells.[22] As noted above, Wells was a convinced Darwinian who learned his biology directly from T. H. Huxley, so it's not surprising to find his fictitious hero Wedderburn discussing Darwin's work in "Strange Orchid" (see chapter 7). Wedderburn tells his housekeeper that "there are such queer things about orchids . . . such possibilities of surprises." And he adds that "Darwin studied their fertilisation, and showed that the whole structure of an ordinary orchid flower was contrived in order that moths might carry the pollen from plant to plant." He goes on to claim that Darwin also noted, "There are lots of orchids known the flower of which cannot possibly be used for fertilisation in that way." He explains that the plants spread asexually, by sending out runners, but "the puzzle is, what are the flowers for?"

Wedderburn wonders whether "*my* orchid may be something extraordinary," in which case he will study it, having "often thought of making researches as Darwin did." Of course, we never see him make those researches, but there's something interesting about the story's conclusion. The day after Wedderburn's near-fatal encounter, the orchid is still in the greenhouse "black now and putrescent," but Wedderburn himself is "bright and garrulous" as a result of his strange adventure. The encounter seems to have proved fatal to the flower but oddly invigorating to the man, an outcome that curiously inverts the encounter between a short-lived insect like a moth and a newly pollinated orchid, that will thrive and spread after the insect has done its work.[23]

The British writer John Collier (who wrote the original screenplay for John Huston's movie, *The African Queen*) also produced a number of science fiction novels with evolutionary and Wellsian themes. Among his short stories was "Green Thoughts" (1931), that was based on Wells' "Strange Orchid," but which also hints, albeit very tentatively, at sexual contact between orchids and humans.[24] Collier's story describes a typically Wellsian "little man" very like Wedderburn; Mr. Mannering is a bachelor orchid collector who also lives with his female cousin and buys an unknown orchid at auction, "part of the effects of his friend who had come by a lonely and mysterious death on the expedition." And, "even in its dry, brown, dormant root state, this orchid had a certain sinister quality. Its gnarled surface resembles a grotesquely whiskered grinning face." If only Mannering had read Wells' story, he would have known what to expect, and would have listened to his cousin Jane's oft-repeated warning that "no good would come of his preoccupation with 'those unnatural flowers.'" But of course he plants the orchid, which soon grows huge and proves carnivorous. First the cat disappears, then cousin Jane, and Mannering was astonished to see that when the first of the orchid's larger buds opens, its flower is "an exact replica of cousin Jane's lost cat. The similitude was so exact, so lifelike, that Mr Mannering's first movement, after the fifteen minutes [of staring], was to seize his bath-towel robe and draw it about him, for he was a modest man and the cat, though bought for a Tom, had proved to be quite the reverse." He is so preoccupied that he doesn't realize the orchid's tendrils are twining around his feet and after "a few weak cries" he sinks to the ground. When he regains consciousness, it's a distinctly vegetable consciousness; he's been absorbed by the orchid and is now a bud alongside a bud of cousin Jane. The orchid Mannering retains enough cognitive power to regard with horror the thought that he might now be named and classified, become the subject of a scientific paper, "possibly even of comment and criticism in the lay press." For "like all orchid collectors, he was excessively shy and sensitive." The prospect brings him "to the verge of wilting," but he's curiously reinvigorated by a bee crawling across his face and popping into what was once his mouth. However, there's a richly comic moment when the bee promptly flies over to cousin Jane and repeats the process. Unfortunately, he has explained the role of insects in orchid pollination to her in detail and can only watch in horror as "he saw the soft and orderly petals of his cousin's face ruffle and incarnadine with rage and embarrassment." Unable to transmit his apology, Mannering summons up "all the chivalry proper to an orchid-collector" and tries to compose his own petals into an expression that will convey "grief, manly contrition, helplessness in the face of fate, willingness to make all honourable amends, all suffused with the light of a vague but solacing optimism; but it was all in vain." All he can manage is a slight flutter of his left eyelid.

The comic possibilities of Wells' story prompted Arthur C. Clarke (best known for *2001: A Space Odyssey*) to write his own parodic tribute to Wells' original, "The Reluctant Orchid" (originally in *Satellite Science Fiction* magazine, 1956), but it's rather less successful than Collier's.[25] Clarke's "little man" is the extremely inaptly named Hercules Keating, a 97-pound weakling whose peaceful orchid growing is constantly spoiled by the intrusions of his aunt Henrietta, a massive, self-confident woman who "usually wore a rather loud line in Harris tweeds, drove a Jaguar with reckless skill, and chain-smoked cigars. Her parents had set their hearts on a boy and had never been able to decide whether or not their wish had been granted." She's a successful breeder of large dogs and "rightly despising men as the weaker sex, had never married." She takes an "avuncular (yes, that is definitely the right word) interest in Hercules," perhaps because her cowering nephew "bolstered up her feelings of superiority" over men. The description of Henrietta spells out, as explicitly as was possible in a fifties science fiction magazine, that she's a lesbian, and she inspires terror and revulsion in her pint-sized, bachelor nephew. Inevitably, Hercules acquires from a dealer an unknown orchid that has come from "somewhere in the Amazon region." An unprepossessing brown lump, about the size of a man's fist, "it was redolent of decay, and there was even the faintest hint of a rank, carrion smell." And equally predictably, it grows into a massive, aggressive carnivorous plant that the orchid-grower decides is the perfect tool with which to rid himself of his intrusive aunt. He starves the plant for weeks before letting her see it, but confronted with the aunt's Amazonian figure, "the plant clutched them [its tendrils] tightly, protectively, *around itself*—and at the same time it gave a high-pitched scream of pure terror. In a moment of sickening disillusionment,

Hercules realized the awful truth. His orchid was an utter coward." The narrator of Clarke's unfunny, misogynist tale concludes by telling his listeners that not only has Hercules not freed himself from his aunt, "he seems to have sunk into a kind of vegetable sloth . . . every day he becomes more and more like an orchid himself. The harmless variety, of course."

When Wells published the progenitor of these orchid stories in 1894, the idea of interspecies mating, especially between plants and animals, was still firmly in the realms of fantasy. However, within a few decades the news of orchids' almost unbelievable "Machiavellian cunning" was spreading to the wider public, and the bizarre fact of pseudocopulation added a fresh dimension to the rich variety of ways in which orchids could be imagined. Just a few years after Clarke's story appeared, the magazine *Science Fantasy* published "Prima Belladonna," one of the first science fiction stories by the British writer J. G. Ballard (who would go on to become world famous as the author of novels such as *Empire of the Sun*, 1984).[26] "Prima Belladonna," however, was a very different kind of tale. Ballard's story is set on a resort world called Vermillion Sands and narrated by Steve, the slightly lackadaisical owner of a flower shop (Parker's Choro-Flora). Steve's flowers were bred not for their appearance, but for their musical abilities (his stock includes azalea trios, and a delicate soprano mimosa). Into Steve's life comes Jane Ciracylides, a stunningly beautiful woman, "even if her genetic background was a little mixed"; the gossips of Vermillion Sands decide there must be "a good deal of mutant in her," not least because she has "what looked like insects for eyes." She's a "speciality singer," whose voice is so extraordinary that it causes her audience to hallucinate, having intense personal visions; one of her first acts in the story is to "sing" a completely real image of an Emperor Scorpion at Steve and his friends whom she catches ogling her.

Although she flirts with Steve and his friends, it soon becomes apparent that Ciracylides is not really interested in them, but in the rarest, most valuable plant in Parker's shop—a "Khan-Arachnid orchid," one of only about a dozen "true Arachnids in captivity." It's a temperamental flower, "a difficult bloom, with a normal full range of twenty-four octaves," but "unless it got a lot of exercise it tended to relapse into neurotic minor key transpositions which were the devil to break." Its behavior creates additional problems because it is the "senior bloom in the shop," so its moods affect all the other plants. As we might expect, it turns out the first Arachnid orchid came from a South American jungle (the Guiana

forest) and was named after the Khan-Arachnid spider, its pollinator. The botanist who discovered it believed that the flower's vibrations not only guided spi-
ders to the plant, but actually hypnotized them.

Ciracylides comes to Steve's shop to look for a flower but the orchid is the
only one that really holds her attention:

> "How beautiful it is," she said, gazing at the rich yellow and purple leaves hang-
> ing from the scarlet-ribbed vibro-calyx.
>
> I followed her across the floor and switched on the Arachnid's audio so
> that she could hear it. Immediately the plant came to life. The leaves stiffened
> and filled with colour and the calyx inflated, its ribs sprung tautly. A few sharp
> disconnected notes spat out.
>
> "Beautiful, but evil," I said.

The woman denies this, asserting that the plant is merely proud. We might
initially assume we're about to meet yet another seductive, feminine orchid,
but this orchid is clearly masculine. Under Ciracylides' influence, the plant
starts to really sing (for the first time in Steve's experience) and does so in "a
full baritone." Steve watches Ciracylides "staring intently at the plant, her skin
aflame, the insects in her eyes writhing insanely" and the Khan-Arachnid seems
equally entranced; when it stretches out toward her Ballard's description of the
flower — "calyx erect, leaves like blood-red sabres" — emphasizes its masculinity.
And when his friends are raving to Steve about her performance at the Vermillion Sands Casino, he tells them it was nothing compared to the one she gave in
the shop. "That Arachnid went insane. I'm sure it wanted to kill her," he comments, to which one of them responds, "If you ask me it's in an advanced state
of rut. Why should it want to kill her?"

Under the orchid's influence, the plants in Steve's shop seem to be going
crazy ("By the third day after Jane's arrival I'd lost two hundred dollars' worth
of Beethoven and more Mendelssohn and Schubert than I could bear to think
about"), until the mysterious mutant singer offers to help and proves to be a
natural at getting the plants to sing in harmony. He offers her a job, joking that
he would treat her just like one of the plants ("I'll fit you out with a large cool
tank and all the chlorine you can breathe").

Pretty soon Jane and Steve are having an affair and become inseparable, but one night—when she ought to be onstage—Steve discovers she has crept back to the shop for a duet with the Arachnid. When he walks in on them he finds this scene:

> The Arachnid had grown to three times its size. It towered nine feet high out of the shattered lid of the control tank, leaves tumid and inflamed, its calyx as large as a bucket, raging insanely.
>
> Arched forward into it, her head thrown back, was Jane.

When he pulls her away, he notices that "in her eyes, fleetingly, was a look of shame." He leaves her in the shop, but when the music ends, he finds Jane has gone. The orchid shrinks back to its normal size and, the next day (in a faint reminder of Wells' strange orchid), the Khan-Arachnid is dead. Precisely what the woman and the orchid were doing is not spelled out, but the scene is unmistakably sexual. Yet while the other *femmes fatales* we've met used orchids as a means to an end, or occasionally used men as a means to orchids, we now have an erotically charged liaison between two unrelated species. (The title of Ballard's story also alludes to the *femme fatale* theme; in Britain the poisonous plant Deadly Nightshade has been known as "Belladonna" since at least the sixteenth century, when Gerard called it by its Italian name, in his *Herball*.) The connection between Ciracylides and the orchid is not unlike those that had been discovered between wasps and orchids, except that in Ballard's vividly imagined encounter, the animal appears to triumph over the plant (but then Ciracylides' species is ambiguous; she's a mutant with insect-like eyes and a name that sounds like the scientific one of some unknown organism). Steve concludes his narration by telling the reader that he never saw the woman again, but "I heard recently that someone very like her was doing the night clubs this side out of Pernambuco." (That final choice of location is unlikely to be accidental—as we saw earlier, Pernambuco is where Swainson actually collected *Cattleya labiata*—and, like several other details in the story, suggests Ballard knew all about orchids and their history.)

Along with their potentially deadly sexuality, the orchid's nineteenth-century association with decadence persists into twentieth-century science fiction. The scent of orchids, so often associated with their supposed sexual connotations, is also the scent of decay; a hint of rotting carrion is often the first sign that we're dealing with a malignant orchid. "Prima Belladonna" is set in the resort of Vermillion Sands, which Ballard was to revisit in later stories, gradually developing it into the ultimate embodiment of decadent pleasure, but one that seemed to be in terminal decline (or perhaps it was designed that way?), a technologically advanced Xanadu in urgent need of maintenance. Moreover, the first story takes place during "the Recess, that long slump of boredom, lethargy and high summer that carried us all so blissfully through ten unforgettable years." It sounds like the kind of place where a latter-day Oscar Wilde would feel right at home, and exactly where one would expect to find a madly rutting, singing orchid whose scarlet-ribbed "vibro-calyx" sounds like something Jean des Esseintes would die to possess.

A similarly decadent fin-de-siècle mood runs through a science fiction novel that first appeared in German, just a few years after Ballard's story. *Der Orchideenkäfig* was written by Herbert W. Franke (first published in 1961, but not translated into English, as *The Orchid Cage*, until the early seventies).[27] Franke's novel centers on a waning humanity who, finding all their problems solved, devote themselves to various hi-tech, synthetic entertainments. Exploration of other planets is one such pastime, but since faster-than-light travel is impossible, something called the "synchromirror" allows information to be transmitted instantaneously across the galaxy. The signal can be used to stimulate the creation of "pseudo-bodies" on distant planets, which humans can control and inhabit, vicariously sensing everything their distant counterpart experiences. Every world humanity discovers is dead; it seems that life is a rare, fleeting occurrence that invariably wipes itself out as soon as it becomes intelligent. We seem to be the only species to have survived past that step in our evolution. So, to add some excitement to this harmless (and rather depressing) pastime, the travelers compete to be the first to explore a new world and discover what its highest life form looked like.

The Orchid Cage concerns one such competition to the seemingly dead cities of a remote sterile world in an almost empty corner of the galaxy. The defensive responses of the intelligent automatic machines suggest that perhaps the planet is not quite dead, maybe the original inhabitants survive somewhere deep underground. As they explore, the humans find abandoned homes, with entertainment chairs that project illusions directly to the nervous system; the absent inhabitants were clearly somewhat like the visitors, free from any need to work or struggle, devoted to sophisticated pleasures. Gradually, the leader of one of the competing teams, Al, admits he has lost interest in the challenge. He wants to know what has become of the original inhabitants, "because that's what will become of us." He proposes they should ignore the Regulations that govern such competitions in an apparently unprecedented situation: "We on Earth thought there was no more work to do, nothing of any interest to find," but it appears they were wrong (92).

As the teams explore the city, they find an impenetrable cover that prevents access to the lower levels, but also recording devices that replay sights, sounds, and even smells as vividly as if the wearer were present. Searching the records, Al watches some kind of solemn ceremony: two of the humanoid creatures who built the machines are setting up some kind of equipment that is partly buried in the ground, watched by an attentive crowd:

> It was not long before there was a movement: something emerged from the holes, yellow shoots, jerking forward unsteadily as if growing. When they had grown to about the size of a human hand one of the shoots grasped at the switchboard and all the other shoots immediately branched off. They grew faster, and again one of them grasped a switch. Leaves sprouted from the stems. Gradually the plant took shape. (108–9)

This plant-like machine was the first of a new generation of sophisticated, self-repairing, self-programming robots that were devised to take care of the alien people's every need. As the machines took over, entertainments were increasingly delivered by direct stimulation of the nervous system and the people's increasingly unnecessary and vulnerable bodies were moved underground to protect them from possible harm. Amid the fleeting images on the recordings is one of cylinders apparently being stored and "inside each was a pink-coloured and fleshy multi-petalled shape." Each container was connected to pipes and tubes and "looked like an orchid in a cage" (114).

Finally, the humans make contact with the guardian machines and are allowed to enter what initially appears to be a "cage of orchids" who are, of course, the people. "For us," said Al, "it is incredible how creatures which were similar to us could have changed into plants," but one of the robot keepers explains that their evolution is continuing, so that the orchids' vestigial stomachs are gradually disappearing since nutrients are fed directly into them. Their brains are wired to computers that allow "them to imagine pleasurable things: peace, happiness, contentment, and other sensations for which there are no words."

They cannot move, communicate, or reproduce because they have no need to. (Exactly how an organism that doesn't reproduce could be evolving is one of a number of rather glaring holes in the plot.) The humans "looked at the flaccid organisms in their protective metal and glass shells, which had in a way attained their goal: paradise, nirvana, everything and nothing in a steamy underground corridor." "So that's what it's like to have no further desires," said Al. He switches off and his pseudo-body collapses lifeless on the alien floor. He finds himself back on earth, sitting in a chair perilously similar to those they had seen on the other planet, with machinery all-too-like that they have observed in the alien recordings. He smashes everything to pieces and leaves the room, going out — perhaps for the first time — "into the open air" (167–74).

Franke's story takes the idea of orchids as an evolutionary pinnacle, the highest of flowers, and combines it with the familiar nineteenth-century idea of degeneration, that an overcivilized race must inevitably decline as it loses abilities it no longer uses. Like other sci-fi texts (beginning with E. M. Forster's pioneering novella *The Machine Stops*, 1909, whose ending is clearly echoed by Franke's), *The Orchid Cage* explores the idea that overreliance on technology will turn us into parasites, becoming simpler as we become dependent on our hosts, the machines we have created. However, the distinctly organic technologies in Franke's story — the plantlike machines and the orchid imagery — help create a different sense of the seductive power of decadent luxury as something that grows up around us and to which we adapt. Faced with a life of unparalleled ease, where would we find the strength to demand a return to a harder life?

In addition to its hints of degeneration, Ballard's story has clear roots in the fin-de-siècle tradition of the orchid as femme fatale. The orchids' associations with luxury and decadence are also used to faintly comic effect in the movie *Barbarella* (directed by Roger Vadim, 1968). When the heroine (played by Jane Fonda) reaches the labyrinth of Sogo, the City of Night, she finds the prisoners eating orchids and is told that "orchids have very little food value and are hard to grow in this climate. It amuses the Great Tyrant to resent the expense of feeding orchids to slaves." Other science fictions make even more unexpected uses of orchids. The hero of "Planting Time," by Pete Adams and Charles Nightingale (1975), is Randy Richmond, the only human aboard a ship on a long "plus-light" trip. This imagined future doesn't include self-confidently sexual women like Barbarella, so Randy — who is apparently typical of space-faring males — whiles away the tedious months of travel by examining "the 3-D pull-out pin-downs from *Stagman* magazine."[28] Yet despite these images and the efforts of the ship's computer (known as a Spacegoer's Companion) to entertain him, poor Randy is deeply bored by the time they land on a planet to replenish the ship's chemical supplies and get some shore leave.

The planet's humanoid inhabitants are "at a fairly primitive stage of development," meaning all contact is forbidden, so the ship lands on an uninhabited island. As he explores, Randy is understandably astonished to discover a beautiful woman wearing "nothing but a short blue shift of some intricately worked material." She never speaks but her gestures are unmistakably inviting:

> She sighed like the murmur of leaves in midsummer and stretched herself out before him, the hem of her garment rising gently to reveal dark and appetising areas of accessibility. Her scent was all cinnamon, musk, and pure violets, stifling rational thought. Randy toppled drunkenly into her and was enfolded by flesh that writhed delicately against his own, and by hair that seemed to caress him with gently powdered tendrils as he plunged and gasped and shook.

Just what the lonely spacegoer was dreaming of, but when he tries to engage his companion in a repeat performance, she is not just uninterested: "her flesh seemed actually to crawl with distaste." But all is not lost; on his way back to the ship, Randy finds a second girl, this time dressed in red but otherwise so similar to the first that they could have been sisters. She is just as stunning and willing as the first, but again, a second bout of lovemaking is repulsed and he finds the girl "as unyielding as a block of wood." A third encounter follows that, like the others, is accompanied by a rich, intoxicating scent, that leaves our amorous hero feeling "as if the planet itself had opened up to swallow him, the grass and the giant leaves closing above his head. The climax seemed to scatter him around the landscape like fragments of a bursting pod." After a night's sleep, the area around the ship is crowded with leafy couches, each supporting an almost identical beautiful and eager female, all beckoning enthusiastically to solicit Randy's attentions. "He ploughed and dug his way across this incredible plantation of sunsoaked skin, discarded garments, and voluptuous welcome, until his responses became too painful to be worth the continuing effort."

So far, so fabulous, at least as far as the lonely spaceman is concerned, but trouble soon appears in this playboy paradise. He finds the first girl again, but she is "unmoving on the stained and fraying couch, her shift draped over her limbs like a rotten shroud." Her skin is now pallid, dull, and sagging; her hair "had coagulated into a limp, repellent mess." Guiltily, he takes her hand to find out what's wrong, and "it parted immediately from the sagging mass of her body and rested soggily in his grip, greenish matter dripping from the severed wrist." Appalled by the thought that he might have infected his playmate with some human virus, Randy flees to the ship to tell the Companion what's happened, but "the computer received Randy's confession in utter contempt." If he had read the data files on the planet, he would have discovered that the planet had been visited before, and botanists had discovered three species of a plant called *Bacchantius* growing there, of which the most unusual was *Gigantiflora*, a member of *Phorusorchidacae*, the local orchid family (the mock-scientific name is not explained but is clearly built from the Latin *phora*, motion). Randy reads the botanical report and discovers that "pollination is by pseudo-copulation, as in many species of plant, but is exceptional in this case in that the pollinating agent is the male *Gaggus*," the planet's primitive humanoids. The flowers have evolved into a reasonable facsimile of female *Gaggus*, and in addition:

> The flowers have a very powerful scent, and while the chemical structure of this has yet to be determined it is known to have pronounced hallucinatory and aphrodisiacal properties, which it is thought acted originally to prevent the *Gaggus* from discovering the true nature of the girl that apparently confronts him. Under the influence of the scent, for example, the male finds the eyes of the plant lifelike and mobile, whereas in reality they are the least successful part of the imitation.

Randy has discovered what it feels like to be one of Pouyanne's wasps, the helplessly deceived pawn of a cunning and highly evolved orchid. However, Randy fights back promptly and the story's ending reveals that he's gone on to become a famed horticulturalist, proprietor of a galactic chain of floral brothels where the "hybrid strains developed by the computer become more delectable year by year."

For much of its early history, science fiction often reflected the rather predictable tastes and values of its predominantly adolescent, male audience, but during the sixties — when flower power joined calls for black power and women's power—science fiction began to develop a broader and more imaginative range of concerns, including an interest in ecology and the impact that humans and their sciences might have on our own and other planets. The famous "Earthrise" photo, taken by the Apollo 8 astronauts in 1968, may have been a contributing factor; the first photograph of the earth from space, that iconic image of a blue marble floating in infinite blackness, provided many people with a vivid sense of how small and vulnerable our planet was. As a result, the picture become something of an emblem for the infant environmental movement (adorning the cover of the "hippy bible," the *Whole Earth Catalog*). Space science and the fate of plants were occasionally linked, notably in the film *Silent Running* (1972), in which spaceships provided the last refuge for the Earth's vegetation.

The various strands of the sixties counterculture provided the context for the most unexpected and remarkable novel to explore the impact of the botanists' discovery of pseudocopulation: John Boyd's *The Pollinators of Eden* (1969).[29] It's the story of Dr. Freda Caron, an exobotanist, who is disappointed when her fiancé, Paul Theaston, does not return from the planet Flora because he needs to finish a study of orchid pollination. By way of apology, he sends her his graduate student Hal Polino and some Florian tulips that Paul has named Caron tulips in her honor. These highly evolved plants have several surprising abilities, including that of not merely making sounds but accurately mimicking those they hear. It's a fairly complex novel, with a great deal of (slightly heavy-handed) satire of scientific and governmental bureaucracy woven into the tale of how Caron and Polino gradually prove that the tulips are intelligent, not only communicating with each other to defend themselves from threats, but able to rapidly alter the behavior of one of Earth's wasp species so that the insects will pollinate them. However, the accomplishments of the tulips are, as we shall see, completely overshadowed by those of the orchids.

The mystery that is keeping Paul on Flora is what pollinates the orchids, given that the planet is completely devoid of insects. He has set all kinds of traps for birds and animals but caught nothing. Whatever the pollinators are, they must be able to see the beautiful flowers, and have a sense of smell, since the orchids "exude a perfume so enchanting that if I could bottle it and ship it home it would devastate the ecology of earth in nine months." As he investigates, Paul begins to feel as if the orchids are "deliberately concealing their secrets from me" (25).

Freda finally gets to Flora to find out what's become of her erstwhile fiancé;

DELL
6996
75c

A WOMAN WITH PROBLEMS . . .
A PLANET WITH SEX APPEAL . . .
A BRILLIANTLY BIZARRE NOVEL
OF OUTER SPACE SENSUALITY

THE POLLINATORS OF EDEN ·

JOHN BOYD

Figure 42. The discovery of pseudocopulation would inspire the
science fiction writer John Boyd to imagine interstellar orchids who
have "found the ideal animal for their purpose"–human beings.
(Cover illustration by Paul Lehr, from *The Pollinators of Eden,* 1970 edition)
Credit: Author's collection.

she finds him on the planet's only tropical island, surrounded by orchids that, like other Florian plants, have evolved completely separate sexes instead of hermaphroditic flowers. The orchids are huge:

> He lifted her to his shoulder for a closer view of the bloom, its calyx petals folding outwards to three feet in diameter. Delicate traceries of red on the corolla petals and the lip gave it the appearance of a *Cymbidium alexanderi*, save for the fact that there was a single stamen with no stigma. Its stamen was almost six inches long (186–87).

Meanwhile, he has solved the mystery of how the orchids are pollinated; large pig-like animals are responsible, but when their numbers get too large and they start to damage the orchid groves, the mobile, intelligent, and carnivorous orchids cull their pollinators to keep them in check (the males, being the larger and stronger plants, form a protective barrier around the orchids' territory).

Paul believes there is now the prospect of a "true symbiosis" instead of the "ecological cold war" with the pig-like creatures; and Freda discovers what he means on her first night when she has a disturbingly vivid dream of being carried to an altar where she's bound and offered "breech first" to a priest who is an orchid, apparently to be killed. But there's no death agony, only a highly sexual dream of being a rodeo rider on a bucking bronco. In the morning, when she accuses Paul of abandoning her in the night, he says "you had them" and gestures at the orchids. She replies, "I dreamed they had me . . . your friends were delightful." Encouraged by her reply, he takes her to meet "Susy," his favorite among the female orchids, who clearly likes Freda too. With Paul's encouragement, Susy embraces her and Freda soon finds the orchid's "lower tendrils sliding between her thighs," "kissing her lightly at first and then more amorously, finding zones of maximum response." After what may be the first (possibly the only) explicit scene of lesbian sex with an orchid, Paul tells her that Flora's orchids have finally "found the ideal animal for their purpose. We are the pollinators of Eden." By mating first with a male then a female orchid Freda has become the orchid's insect, rewarded for her pollinating efforts not with nectar but with intense sexual pleasure. Apparently the orchids can both sense and stimulate human desire (190–99).

Like other science fiction orchid stories, Boyd's *Pollinators of Eden* contains a hint of decadence and decay; Flora orbits an ancient sun that is near the end of its life, evolution has run its course, and no future progress can be expected (as in Franke's novel, the orchids represent some ultimate end point of evolution). There are hints that the Earth is also facing its finale, that the whole universe has reached its point of maximum expansion and is about to collapse back in on itself. The novel includes some complex political maneuvering back on Earth over whether or not Flora should be colonized. The space-going navy are concerned that several of their men have already disappeared into the flowerbeds, shedding their clothes and inhibitions, never to be seen again, so they bribe the chair of the Senate committee that's hearing the colonization petition. He uses his casting vote to declare Flora a pariah planet, unfit for human habitation. The chairman condemns those who want to colonize Flora as "Lotus Eaters," and vows to prevent any "Tahitis in space" or "cul de sacs of flowers," which he asserts can only rot "the moral fibre of our civilisation"—humanity needs struggle. (There are obvious echoes here of the fin-de-siècle fear of imperial degeneration, that white men might be corrupted by the ease and sensuous temptations of the tropics and "go native," that lurk beneath the surface of some of the late-Victorian orchid adventures.[30]) "Man could not have walked so splendid in the sunlight if he had not been cast out of Eden," the chairman declaims. When hecklers condemn him as a heretic and call on him to die, he responds, "Die I shall; but I shall die standing on the legs of a man, hurling curses at the dying sun, and not as a vegetable, succumbing silently to the first nips of frost" (88–89). No orchid cage awaits us if this cantankerous politician gets his way. Meanwhile, others are pursing colonization because they're convinced that the mating of humans and orchids may produce seeds so resistant that they could survive the collapse not only of planets, but also of the universe itself; if they were shot outside space and time they might smuggle human DNA into a future universe, as yet unborn, giving our next attempt at evolution a head start. However scientifically implausible this is, Flora offers a hope that is far more than escapist lotus eating. Although *The Pollinators of Eden* can be criticized as a novel, there's no disputing its imaginative scope and the scale of its ambitions.

In the imaginations of science fiction writers, the various meanings of orchids became ever-more vivid. These extravagant tales clearly had their origins in the orchidmania of the nineteenth century; the now-familiar paraphernalia of steaming jungles and deadly native females provided many of the themes that first animated science fiction's orchids. The idea of killer orchids also lingered on into the twentieth century; in the most famous of all deadly plant stories, John

Wyndham's *Day of the Triffids* (1951), the fatal plants were originally the product of experimentation by Soviet biologists trying to create an alternative source of multipurpose oil. Near the beginning of the book, a mysterious character called Umberto offers to sell the secret of the new oil that threatens to put Western producers out of business; when questioned, he identifies orchids as one of the monstrous plant's components and the Triffids produce clouds of minute, airborne seeds, a clear sign of their orchid ancestry.[31]

Once orchids had been liberated from any vestige of earthbound plausibility, their sexiness could take exuberant flight. Nevertheless, Boyd retained a clear sense of where his story had begun; Flora's male orchids have forked roots that are endowed "tubular appendages which reveal that the ancients used scientific precision when they named the plants 'orchis.'" (19). That ecological accident, the presence of paired tubers in the first species that Western science investigated, had begun a long and remarkable chain of associations, whose imaginative unfolding the ancient Greeks could scarcely have begun to imagine. Not only were the sex lives of orchids, insects, and people becoming interwoven, but one of Western Europe's most ancient and unquestioned assumptions was beginning to be seriously questioned: orchids, like everything else in nature, were no longer simply a resource to be exploited. *The Pollinators of Eden* dramatizes a real question: once we realize that plants possess strategies and goals to ensure their survival, will we cooperate with the plants, or thwart them?

11

Endangered Orchids

On August 23, 1850, the British botanist Joseph Hooker, then traveling in India, wrote to his father, William, who was director of the Royal
Botanic Gardens at Kew. Joseph was dismayed by the scale of orchid
collecting he was witnessing, and described how collectors from the
Calcutta Botanic Gardens had just filled 1,000 baskets with them: "the
roads here are becoming stripped like the Penang jungles, and I as-
sure you for miles it sometimes looks as if a gale had strewed the road
with rotten branches and Orchideae." As a result, he had decided not
to spend any time or money on orchids, because "the only chance of
novelty is in the deadly jungles."[1] The thought that some orchid species
might be pushed to extinction by such excessive collecting doesn't seem
to have occurred to Hooker, who was soon breaking his resolution not
to collect orchids and gathered 360 plants of a striking blue orchid,
Vanda coerulea. As he noted, it took seven men to carry the baskets "but
owing to unavoidable accidents and difficulties, few specimens reached
England alive." A rival collector got a single man's load to Britain where
it sold for £3,000 (over £2.2 million/$3.5 million today), with indi-
vidual plants fetching as much as £10.[2]

At midcentury, few seemed concerned about the orchids' future,
but as late-Victorian orchidmania took off, some enthusiasts started

to worry about the survival of their favorite plants. Benjamin Williams, whose "Orchids for the Million" series did so much to make the flowers popular (see chapter 4), lamented in his *Orchid Grower's Manual* that "hundreds of beautiful species" that had cost so much time and effort to collect, had "rapidly died out" in Britain because nobody knew how to grow them. He advised against collecting on too large a scale and told collectors to ensure they "leave some for stock in their native country, instead of sacrificing the whole produce of a district." As Williams noted, there were several species of orchids cultivated in Britain "of which there has only been one importation." Some were very desirable, but "although our collectors have been in search of these scarce plants, they have not been successful in again finding them." He urged collectors to focus their energies on discovering how to get a few plants home alive instead of stripping whole areas in the hope that a lucky handful would make it. "The present destructive system, or want of system, leads to a loss of capital, and is, besides, an annoyance to both the sellers and purchasers." And yet some collectors "seem determined to exterminate certain kinds of Orchids from their natural localities, without anyone deriving benefit thereby" (see chapter 6).[3]

Any suspicion that Williams might have been exaggerating the threat posed by orchid collecting are dispelled when one reads the memoirs of a typical orchid collector, Albert Millican, who collected in the Northern Andes. He made five journeys to the "orchid districts" of South America between 1887 and 1891, where he claimed to have met everyone from "the wild Indian of the forest to the polished and educated senator of the court."[4] *Cattleyas*, especially the spectacular *Cattleya mendelii*, were a particular focus but when he first started exploring Columbia, northeast of Bogota, he found that "not a single one is to be found within many days journey from here on mules." He climbed high into the mountains where the condors nest and recorded that

> when the first plant-hunter arrived, these dizzy heights offered no obstacle to his determination to plunder. Natives were let down by means of ropes, and by the same ropes the plants were hauled up in thousands, and when I visited the place all that I could see of its former beauty and wealth of plants was an occasional straggling bulb hung as if in mid-air on some point only accessible to the eagles (114–15).

Many of the rare orchids Millican wanted, such as the spectacular *Odonto-glossum crispum*, only grew high on trees, so when he finally found some he blithely asserted that the jungle grew so quickly that "cutting down a few thousands of trees is no serious injury; so I provided my natives with axes and started them out on the work of cutting down all trees containing valuable orchids." At first they chopped indiscriminately, but

> soon became adepts at plant-collecting and would bring to our camp several
> hundreds of plants each night, with occasionally a few *Odontoglossum odo-*
> *ratum* and *Odontoglossum Corodinei* mixed amongst them. After about two
> months' work we had secured about ten thousand plants, cutting down to
> obtain these some four thousand trees, moving our camp as the plants became
> exhausted in the vicinity (150–51).

From a twenty-first century perspective it is hard to read of such wholesale ecological vandalism without weeping; four thousand mature rain-forest trees cut down and left to rot in order to collect orchids, most of which would die before they reached Europe, just so that wealthy collectors could adorn their greenhouses with overpriced rarities. And *Odontoglossum crispum* is notoriously hard to grow, even today, so of the few that made it to Britain, most probably died soon after. Less than a decade later, Frederick Boyle (coauthor of *The Orchid Seekers*) reported that *O. crispum* had all but disappeared thanks to the depredations of collectors.[5]

Millican seemed oblivious to the damage he was doing; just a few pages after he describes the swath he cut through the forest, he mentioned finding the grave of one of James Veitch's collectors, who served the firm so well, "long before the wholesale plunder and extermination of the plants brought about by modern collectors" (163). And, unsurprisingly, Millican is no more sympathetic to the indigenous peoples than he is to their habitat; he and his party were attacked by "Indians" with poisoned arrows and one of Millican's group died. They were forced to retreat and Millican regretted that "the orchid gems of the headwaters of the Opon" would remain untouched "until the last red man has disappeared from his territory" (170).

Fragile Specialists

While overcollecting certainly had an impact on many orchid species in the late-nineteenth century, the rarity of many orchids is partly explained by the biology of orchids themselves. In the 1920s, the Harvard orchid biologist Oakes Ames (see chapter 8) was one of many who described orchids as the mostly highly evolved family of flowers, with very complex and specialized flowers. He was therefore surprised that some botanists regard them as a "decadent group":

> The outstanding characteristic of the Orchids that is regarded as a sign of decadence is lavish yield of seed coupled with sparse distribution. This seems to indicate a waning capacity to compete with more vigorous plants.

The production of so many tiny seeds combined with so many individual orchid species becoming rare was interpreted as a sign that orchids were on their way out, too specialized to adapt so that their place in nature was becoming ever more precarious. This might seem like a meaning of "decadence" rather different from the one used by poets and playwrights, but the meanings were closer than one might expect. According to Ames, orchids

> have been mentioned by a recent writer as occupying the same position in the floral world that flamboyant motifs hold in the development of a national art. Just as flamboyancy in art has marked the passing of a civilization, so in the animal and vegetable kingdoms an excessive development of iridescence and fantastic excrescences indicates types of development that precede decadence and extinction.[6]

The orchid empire was, like the Roman one, declining and about to fall—decadence being the undoing of each. There's even a suggestion in the unnamed writer's comments that the "excessive" flowers of the orchids were like the Ancient Roman orgies, evidence of sexuality run amok because no longer properly directed at reproduction. At the very end of the nineteenth century, a French botanist, Noel Bernard, discovered that orchids depend on their roots forming a symbiotic relationship with a fungus, called mycorrhizae (the plural of mycorrhiza).[7] This relationship was also offered by some as further evidence of decadence; like the Ancient Romans, the orchids keep slaves, which encourages them to become dependent and weak, where once they were martial and manly.

The orchids' mycorrhizal so-called slaves form when various kinds of fungi that occur naturally in the soil form a symbiosis with a plant; without them, orchid seeds cannot germinate (this was why they had been so hard for nurseries to grow). They are vital because orchids have the smallest seeds of any plants, weighing as little as one-millionth of a gram. That's a huge advantage to orchids in some ways. Each seed requires very little biological energy to make. In biological terms, they are "cheap" and so can be mass produced and spread widely. And because they are so light, the wind can carry them for many hundreds of miles. (Genetic testing has shown that some isolated populations of rare Japanese orchids are so closely related to mainland Chinese orchids that they must be descended from tiny seeds that traveled as much as a thousand miles, across hundreds of miles of open ocean, to colonize the island. The orchids of Tahiti have made an even longer journey, from New Guinea, almost 4,000 miles away.)

However, minute seeds have their drawbacks; each contains very little stored energy to get the plant growing. The largest seed of any plant is that of the Seychelles Double Coconut (or Coco de Mer, *Lodoicea maldivica*), which can weigh up to 30 kilograms (66 pounds).[8] Each nut contains enough stored energy to get a new palm tree off to a good start. By comparison, an orchid seed must gather energy immediately after germination by growing underground for a long period before it can emerge. Obviously, an underground plant cannot photosynthesize and make food from sunlight, so orchids rely on mycorrhizae to provide nutrients directly to the orchid by breaking down large molecules such as cellulose into simple sugars.[9] As the orchid expert William Stearn jokingly put it, when he was listing the "crimes" committed by orchids:

> From Karl Marx's standpoint, wealthy Victorian orchid-growers enjoyed their orchids as a consequence of the sweated labour of underpaid miners whom they never saw. Research on mycorrhiza suggests that orchids can be regarded in much the same way—as ostentatious floral capitalists dependent upon the unseen obscure activities of fungi.[10]

In reality, as Stearn and his audience would have understood well, mycorrhizae are not "slaves" in any sense; symbiosis is a relationship of mutual benefit. Some orchid seeds are already associated with mycorrhizae when they're dispersed, but many others depend on luck; if the right fungi are present in the soil they land in, they will flourish; if not, they will either never germinate or will fail to thrive.[11] As a result, despite the massive number of seeds many orchids produce, the success rate for germination is often very low, as it must be with any species that produces a lot of small, cheap seeds (or offspring). The poet Alfred Tennyson (who gave us the phrase "Nature red in tooth and claw") lamented that "of fifty seeds" Nature often "brings but one to bear" (*In Memoriam*, 1850). For orchids, it may take 50,000 seeds to produce a single plant.

The orchids' dependence on mycorrhizae was, as Ames noted, sometimes cited as another example of their decadence. The relationship was sometimes referred to in the 1920s as "luxury-symbiosis," to separate it from healthier, more balanced, symbiotic relationships. A curious British biologist, Hermann Reinheimer, described luxury-symbiosis as a relationship "leaving the organism which has most to lose through inferior association, namely the orchid, the poorer in the end. Often enough the moribund condition of the orchid is marked by its sickly appearance [and] . . . its sparse distribution." As a result of their dependence on the fungus, he argued that the roots of orchids are like teeth that have eaten "too much soft and sloppy food," so they decay. This supposedly scientific description of the symbiotic relationship carried more than a hint of moral disapproval; clearly a more hard-working plant ought to be able to stand on its own two feet (or roots), not rely on underground underlings to bring it its food.[12] Once again, the long-standing cultural meanings attached to orchids — their connection to decadence — seems to have influenced the supposedly objective claims of science.

However, Ames was dismissive of all claims that orchids were either in decline or headed for a fall. As evidence that they were perfectly capable of holding their own in the struggle for existence, Ames offered the re-vegetation of Krakatau (or Krakatoa) following the massive volcanic eruption of August 1883. Within 15 years of an explosion that obliterated virtually every living thing on the island, several orchid species (*Vanda sulingii* — now called *Armodorum sulingiii, Cymbidium finlaysonianum*, and three others) were already growing again. He concluded that orchids are "perfectly capable of entering into successful competition and . . . are quite able to hold their own in a free for all combat."

The key to the orchids' success in recolonizing the devastated island was that they were "well adapted to dispersal by the wind"; orchids' tiny seeds may be fragile and lack nutrients, but are light enough to ensure they're among the first to reach any newly available habitat. By 1897, 32% of the newly reestablished flora of Krakatau (17 species) were plants that had blown in from Java, Sumatra, and elsewhere: eight were from the daisy family (*Asteraceae*), five were grasses (*Poaceae*, or *Gramineae*), and four were orchids. When you consider that grasses and the daisy family are among the world's most widespread plant families, and include many of the planet's most successful weeds, orchids are holding their own in pretty rough company.[13]

The orchids of Krakatau perfectly illustrate the paradox of the rare but common orchids; their tiny seeds allow them to spread widely, but make them hard to germinate and become established. They are found almost everywhere, but—as Darwin himself noted—are seldom common enough to be conspicuous. Although there are so many orchid species, many of them are endangered because they are so specialized, dependent on meeting the right mycorrhizal fungus or a specific pollinator; small changes in their environments can quickly leave orchids endangered. These extraordinary plants are therefore, in addition to all their other amazing properties, perfect tools with which to study and measure the impact humans are having on the planet.

The Spider Orchids of Sussex

Professor Mike Hutchings is a well-respected plant ecologist who was for many years executive editor of the *Journal of Ecology*, one of the world's most prestigious ecology journals.[14] He spent his whole career at the University of Sussex, the only British university that's entirely enclosed by a national park, the South Downs. The university's 1960s red-brick campus is surrounded by chalk downland, the same landscape over which Charles Darwin and Jocelyn Brooke walked, recording and collecting orchids.

Mike grew up close to the East End of London, near Ilford, where trees and open ground were in short supply; he never saw anything as exotic as an orchid while he was growing up. Early in his academic career, one of his mentors advised him to pursue a dual research strategy: to mix relatively short-term projects that would provide a steady stream of results and publications, with much longer-term ones that could provide very different perspectives on plants. Orchids, for Mike, were "the casserole on the back burner" of his research career, a slow-cooking project that he kept coming back to during breaks between other activities.

In the early 1970s, when Mike began his career, ecology was dominated by plant dynamics, studies of the rise and fall in the number of individuals within a plant population. Although such studies were a useful step on the way to today's more sophisticated ecological understanding, they seem pretty crude and uninteresting now. They were often purely descriptive, charting the numbers of plants and graphing whether they rose or fell over time. Mike was part of a generation of ecologists who rapidly became dissatisfied with such work—it seemed too like the old-fashioned, descriptive natural history of the nineteenth century. He and his contemporaries began to rethink the ways plant population data could be analyzed, looking for deeper patterns and searching for causal explanations as to *why* particular changes were happening.

Exploring these issues became the focus of Mike's major orchid study. In 1975, he marked out twenty square meters of chalk downland at a nature reserve near Brighton and just a few miles from his university. By measuring distances from his newly planted corner posts, Mike plotted the position of every early spider orchid (*Ophrys sphegodes*) that was growing in the plot that year. The species was endangered, so conserving it required a detailed understanding of exactly how many were growing and the precise factors affecting its survival. Each individual plant was measured and meticulous records were kept: whether or not they flowered each year; exactly how tall the flower spike was; and, how many flowers and leaves were produced. The orchid data were correlated with information about the other plant species in the plot, together with detailed data about the weather, including rainfall and temperature.

Mike returned to his plot every May when the orchids flowered, year after year, working on his hands and knees, going over every square inch of the plot. Spider orchids are hard to spot; when Mike took me to see them, I walked past dozens before I saw a single one. When they're not in bloom, they may produce only one leaf, which may be no bigger than a fingernail, and even when they flower, they are no more than 12 centimeters (4½ inches) high, with chocolatey brown flowers that are hard to spot. Yet, Mike and his students persisted every spring, mapping and recording each individual plant, even those that didn't flower.

Figure 43. Mike Hutchings on the South Downs near
Brighton with one of his beloved Spider Orchids.
Credit: Jim Endersby.

It soon became clear, from the information collected, that it would be vital to keep collecting data for many years, because each individual year's data were full of unexpected uncertainty. First, there was the problem of finding out whether any new orchids were being added to the population, assessing "recruitment," as ecologists put it. Because of their tiny seeds (see chapter 8), orchids grow underground for some time, in partnership with their mycorrhizae, before they have accumulated sufficient energy to emerge into the light. The underground life of the newly germinated orchid seed presents a problem for orchid ecologists and conservationists because it takes several years to find out how many new seeds

Figure 44. One of the tiny, inconspicuous early Spider Orchids
(*Ophrys sphegodes*) in the author's hand.
Credit: Jim Endersby.

have successfully germinated and survived long enough to emerge as new plants in the plot of ground you are studying. Unless recruitment exceeds mortality, the population is doomed to extinction.

The problems of conducting a census of orchid plants are further compounded at the other end of an orchid's life; it's very difficult to be certain when a given plant has died. If the plant at a specific grid point produces no leaves or flowers you might reasonably assume its life was over, but spider orchids — like many other plants — have tubers and as a result, they can remain dormant for a year, or two, or several. Indeed, some species seem to be able to survive a decade or more of dormancy and then spring back into activity. Near Mike's orchid site was a golf course that included some ancient ruins, long neglected and covered in many years of dense bramble growth. When it was finally decided to manage the land properly the brambles were cleared and the next spring the site was covered in the sturdy little blooms of the Early Purple orchid (*Orchis mascula*), whose bulbs must have lain dormant under the brambles for many years. After several years of study, a careful analysis of the data revealed that for the Sussex orchids, about half the living plants failed to show up above ground each year.

Early spider orchids are relatively short-lived. From the year in which they first show above ground, these orchids have a half-life of just 2½ years (i.e., the length of time it takes for half the population to die off). So, constant replacement of dying plants is crucial for the population to survive. Mike's study revealed that the way humans manage the land where orchids grow is also crucial. The South Downs is one of the oldest human-made landscapes in Britain, the product of forest clearance that made way for centuries of subsequent farming. The traditional livestock for the chalk downland has always been sheep; the Southdown breed (developed in the later eighteenth century by John Ellman of Glynde) is named after the Downs. However, in the mid-1970s, European subsidies made cattle a more commercially enticing choice for Sussex farmers and the flocks of sheep began to be replaced by cattle herds. For the orchids, the cows were a disaster. Grazing sheep nibble the vegetation, clipping the grass like woolly lawnmowers but leaving the roots and bulbs of plants (including those of orchids) untouched and able to re-sprout. By contrast, cows wrap their thick tongues around a clump of plants and tear the whole lot up by the roots, often getting a mouthful of soil with their meal. Being much heavier than sheep they are more likely to slip when the steep slopes of the Downs are wet. The downland topsoil forms the thinnest of layers and a sliding cow soon strips it away, exposing the chalk below and crushing fleshy underground plant organs, including orchid tubers. The cows began to destroy the thousand-year-old ecosystem of the Downs, putting the orchids in severe jeopardy. Fortunately, when the subsidies changed in the late seventies the sheep staged a comeback, but careful management is still needed to ensure that the orchids survive. Livestock need to be kept away from orchid-rich areas from flowering time until seed has been set; if the sheep are allowed to graze all year round, they are almost as damaging to the orchids as the cows.

Given the difficulties involved in accurately measuring the orchid population's size over the years, it is perhaps not surprising that after nine years of annual observations, Mike was ready to quit; May is a busy time for academic plant biologists, with lectures to give, exams to set and mark, and student fieldwork to supervise. Mike also sings in the Brighton Festival Chorus, and May is festival time, when concert preparation is at its peak. But he was persuaded that to stop after nine years was silly; he should at least go on until he had ten years of data. Before the tenth year was up, Mike had obtained some additional funding to support the work, so the project continued and—thanks to further, small infusions of money—it carried on for 32 years, making it possibly the longest-running ecological study of any plant species. Certainly (at least at the time of writing), it is the longest-running study of any species of orchid.

Generally speaking, orchids are not especially fragile or exotic plants (they could hardly have evolved into one of the world's largest families of flowering plants if they were). Most of the British native orchids are rather weed-like in their behavior. Because they produce great clouds of tiny wind-blown seeds, orchids can quickly colonize newly disturbed ground (they often pop up alongside new roads, sometimes carpeting the ground). Similarly, the right pattern of grazing in downland opens up spaces where orchids can germinate, just as a fallen tree in the rainforest can let in enough light for new seedlings to emerge. However, as the complex relationship between sheep, farmers, and European subsidies reveals, part of the orchids' success is that they have evolved to take advantage of very specific niches, including those created by human agriculture. The workings of government bureaucracy may seem glacially slow to us, but the pace of change is frighteningly fast by evolutionary standards and a change in the pattern of subsidy can change the orchid's habitat much faster than the little plants can evolve. For this reason, Mike believes that involvement in the politics of conservation cannot be avoided by plant ecologists. Some academic ecologists prefer to maintain a distance between their work and its political implications, and are liable to get irate when the public confuses the scientific study of ecology with the campaigns of environmentalism and conservation, but for Mike, there's no avoiding the wider issues. He sees academics as having an obligation to the public (who ultimately pay for much of the research done in British universities). Mike believes it is part of an academic's role to explain why he has spent 32 years counting orchids, what the point of this type of research is, and the significance of the insights gained from it.

Long-term research is hard to do. Apart from the persistence needed, it's hard to get funding. Academic funding is increasingly focused towards short projects; running a study over three decades instead of the more typical three years has provided unexpected and invaluable insights. For example, the Sussex orchid study revealed that the orchids' flowering time varied each year, as one would expect. Some years were wet, others dry, some had more frost, others more sun. Over two or three years, no pattern is discernible. But over 32 years, clearer pictures emerge. On average, Britain's spring is getting warmer as the years pass and for every 1°C of extra warmth in any spring, the orchids bloom six days earlier. Despite variations between individual years, the three decades of study revealed that the orchids were blooming three weeks earlier than they had when the study began.

These results led Mike and some of his colleagues to wonder whether earlier flowering after warmer springs was apparent over longer than the 32 years of his study. They hit on the possibility of examining herbarium specimens to seek a similar effect. A herbarium is a library of dried pressed plants. At first, it's not obvious how long-dead, desiccated specimens could tell us anything about the ecology of living plant communities, yet they can because herbarium specimens are usually collected only when species are in flower and the date on which they were collected is always recorded. In research that continues as I write, Mike and his team are analyzing the relationship between flowering date and historical weather records using herbarium specimens of early spider orchid collected as far back as 1850. The relationships seen in the field study and the herbarium specimens is the same. Not approximately, but exactly: for more than 150 years, the orchids bloomed earlier following warmer springs — an extra six days earlier per degree of greater warmth. Over the past century and a half, the average temperatures in Britain have been rising, with more and more warm springs. The field data provided completely unexpected evidence of the effects of climate change on flowering time, and a way of predicting future changes in response to further rise in temperature. Moreover, the corroboration of the field study results by the herbarium data showed that information about other species stored in many types of biological collections could be used to predict how the timing of events like flowering, flying of insects, and egg-laying would respond to climate change. There are an estimated 2.5 billion dated specimens of plants and animals stored in herbaria and other collections around the world, offering huge potential to predict changes as temperature continues to rise.

The field and herbarium studies showed how the timing of orchid flowering was affected by weather and how it would respond to climate change. But there is another reason why rare species like *Ophrys sphegodes* might be endangered even further by the changing climate. Just like the flowers Pouyanne studied in Algeria almost a century earlier, the Sussex spider orchids are pollinated by bees, specifically by male bees that are fooled into pseudocopulation by the flower's mimicry. Nearly all cross-pollination of the early spider orchid is carried out by males of the solitary mining bee *Andrena nigroaenea*. At the moment, the male bees start to emerge before orchid flowering, and the orchid flowers before many of the female bees are flying. Until females are on the scene, attempted copulation with the orchid is the only outlet for the amatory ambitions of the male bees. Records from field observations and from bee specimens stored in museums show that bees fly earlier after warmer springs, so it might be thought that warm springs present no problems for orchid pollination, but the story is much more complex. It turns out that spring temperature affects the flying time of male bees more than it changes flowering time in the orchid; male bees fly *nine* days earlier per degree of greater spring warmth. So, as the climate warms up, the crucial synchrony between the timing of flowering and the flying of the bees that pollinate the orchid will be lost. On top of this, the flight date of female bees advances by *fifteen* days per degree of greater spring warmth. This means that, as the climate warms, there will be more female bees on the wing when the orchid is flowering. The male bees pollinate the orchids, but only if there are no females around (male bees may be dumb, but they're not that dumb); if female bees are plentiful, the orchids will be neglected. So global warming—the result of humans changing the planet's climate—threatens the delicate relationship between the orchid and its pollinator.

When Darwin predicted with his yet-to-be-discovered long-tongued moth and its orchid, he foresaw that if either were to become extinct the other might also be doomed. We could now be seeing a similar orchid-based catastrophe developing not in distant Madagascar, but ten miles from where I'm writing this book. And it's not because of overzealous Victorian collectors; it's because over the past two hundred years we've been burning fossil fuels at an ever-increasing rate, and have only very recently begun to realize the consequences of our actions.

Conclusion
An Orchid's-Eye View?

This book has its origins in coincidences. I have a long-standing interest in Darwin's botanical work, and a passion for the stories of both Raymond Chandler and H. G. Wells, and almost nothing is better than eating vanilla-flavored chocolate while watching *In the Heat of the Night*. Orchids proved, quite by accident, to be a link between many of my interests. Idle curiosity about these flowers has led me into many years of research and along the way, I found all kinds of unexpected connections. I hadn't considered how my chocolate or its flavoring came to be in the shop where I bought them, much less that they were a chapter in the history of aphrodisiacs (although, sadly, it appears that there is no evidence that vanilla or any other orchid is really an aphrodisiac). I had no idea that there was a genre of "killer orchid" stories, much less that it would shed light on European attitudes to the places and people they were subjugating in the nineteenth century, or to the changing role of women in Victorian society. It had never crossed my mind that Darwin's orchid research would leave me pondering the theological problem of evil. Some of my coincidences have cohered into patterns, revealing strange but (I hope) illuminating connections between apparently unrelated histories—of literature and science, empire and sexuality, cinema and horticulture, and many others.

One of the biggest surprises to me was spider orchids, pseudocopulation, and the research being done at my own university that connected orchids to climate change. This local story is, of course, just one tiny part of a global effort to study and conserve orchids, many of which are currently endangered because of humanity's impact on their habitats. The International Union for Conservation of Nature (IUCN) currently has over 400 orchid species listed as "critically endangered," "endangered," or "vulnerable," and for many more, there is not enough data to know how the plants are doing.[1]

Given that some orchid species are facing extinction, largely thanks to us, it's surprising to realize that many others are now more popular and widespread than they have ever been; more people are growing them in their homes, for example, than even at the height of orchidmania. Their current resurgence began in 1922, at Cornell University in upstate New York, where an American biologist, Lewis Knudson, finally figured out how to get orchid seeds growing. The French botanist Noel Bernard originally discovered that mycorrhizae are essential if orchid seeds are to germinate and his work was followed up by a German, Hans Burgeff, who managed to germinate orchids in the lab, but only if he harvested some of the beneficial fungus from the parent plant. What Knudson realized was what it is that mycorrhizae actually do for orchids; they were actively involved not in the *germination* of orchid seeds, but in feeding them once they had germinated, which is what enables such tiny seeds to survive long enough to put out leaves and begin to photosynthesize.[2] Once Knudson realized that, he found he could give orchids the sugars they needed directly and as a result discovered they could be grown on agar gel, a cheap jelly-like medium derived from seaweed that was used in biology labs to grow bacteria. Within a few years of Knudson's discovery, orchid-growing was completely transformed.

Within a few years of Knudson's discovery, orchids were such a big business that the American commercial magazine *Fortune* commissioned a feature about them. The assignment went to a young journalist, James Agee, who is now celebrated for his documentary account of the lives of sharecroppers during the Depression, *Let Us Now Praise Famous Men* (1941), but was still largely unknown when his orchid article appeared in December 1935.[3] Agee focused on the career of Thomas Young, who had an orchid nursery at Bound Brook, New Jersey, and was the first commercial grower to exploit Knudson's discovery. According to Agee, Young was so successful that he soon couldn't afford to house the number of orchids he was producing. In 1929, he sold the business to a Wall Street investment company, Charles D. Barney & Co., who paid $2,750,000 for Young's business (over $100 million/£62 million today), expanding it into a national network of growers and wholesalers while keeping Young's name. Profits had gone up and down, but—as Agee wryly observed—when one considers that the business was founded at the start of the Great Depression so that its first six years were "six of the most eminently lean years of an era," the fluctuation in profits was "by no means so much as you'd have every reason to expect" given that orchids were "one of the most eminently useless commodities in existence."

Agee gave *Fortune*'s readers a detailed and fascinating account of Young's orchid production line. It began with the orchids being hand-pollinated by humans (presumably it was too hard to get the bees to work regular hours and follow orders), after which they each produced between 500,000 and a million seeds. These were sown into test tubes full of agar, 200 to a tube, using "platinum-pointed, flame-sterilised needles." A crop of thousands of orchids could be set growing on a single workbench. After a year, about 50 of the hardiest shoots (still only 1/4 of an inch, about 6.5 mm, high) were selected and repotted. After another year, the 30 biggest were planted out, five to a pot, and a year after that, the best got individual pots. At each stage "the weaklings are thrown away." After three years, the original 200 seeds had produced 25–35 healthy plants, with almost no losses from disease. It then took four further years for the orchids to flower, and six to produce a full crop. Orchids have to be cut in full flower (their buds won't develop any further after they're cut, unlike some flowers), their waxy petals make them naturally long-lasting, but they were also refrigerated for several hours to "harden" them. After being carefully wrapped in waxed paper (that wouldn't absorb moisture) they were packed in reinforced boxes and labeled "Handle With Care" in "flashy red italics," and then shipped express to wholesalers and finally flower shops, sometimes a thousand miles from where they had been grown. Christmas was one of the peak orchid seasons and in 1934, the famous train called the 20th Century Limited needed an extra car just to carry Young's orchids. A few years later, the *Saturday Evening Post* described the Thomas Young Nurseries as "the General Motors of the orchid business," capable of producing "An orchid a minute" (and, naturally, they mentioned Nero Wolfe and retold a few old stories about hunters for lost orchids who sometimes "leave their own heads in the orchid jungles"). The paper described how Young's 56 greenhouses covered an area equivalent to 62 basketball courts and supplied half the orchids worn in the United States each year — including those seen at the White House. Young's even employed a full-time publicity specialist, June Hamilton Rhodes, whose job was to raise the orchid's profile while protecting its classy image. Thomas Young refused to supply flowers for Edward G. Robinson's *Brother Orchid* (see chapter 9) and "after the publicity stills appeared, one could understand why," since one showed "a cigarette girl, with skirts reaching to her thighs."[4]

Agee didn't choose to write about Thomas Young because of any interest in orchids — he just needed to earn a living from his writing. And despite its celebration of American business ingenuity, Agee's story conveys a sense that he really didn't like orchids. For example, after mentioning how Darwin first unraveled their "genital subtleties," he added "we had best say little more here than this: nature contrived every flower, as you must know, as nothing more nor less than an invitation to the rape," and that orchids take advantage "of that privilege" in a way that is "spectacular and complex far beyond the point of mere abuse." Given how much evolutionary effort flowers have put into attracting insects, "rape" hardly seems an appropriate word (although if Agee had learned how orchids trick their pollinators, perhaps "abuse" would have been forgivable). Agee confessed to a deep dislike of orchids in a private letter. While admitting that "the flower itself isn't responsible" he still found "people's reactions to it have been and are so vile that I hate its very guts along with theirs."[5] In a later letter, he admitted he was "silly" to dislike orchids so much, but did so partly "by transference," because he disliked the kinds of people who liked the flowers. He also felt that "liking a thing because it is the Largest, the Loudest, the Most Expensive, the most supercharged with Eroticism, Glamor, Prestige — I don't like. Automatically thinking a thing is beautiful for such reasons I like even less."

In one of his letters, Agee quoted parts of his manuscript in progress (which, luckily for the orchids, did not make it into the printed version). He had originally written that orchids were "a psychopathic nightmare in technicolor," as attractive as the bare backside of an aroused baboon. When a man gave orchids to a woman they were simply a "conspicuous expense to assert your opinion of her as something pretty nice to be seen with." Easter was the biggest week for orchid sales, and in his draft Agee mocked the use of the flowers "to lend Class to any occasion of social or sentimental stature" such as "the embarrassment and ultimate destruction of Death through the glorious resurrection of Jesus Christ."[6]

Figure 45. The Roosevelts on their way to a Christmas service in 1934. The orchids worn so prominently by Eleanor Roosevelt would have been supplied by Thomas Young Nurseries. Credit: Library of Congress Prints and Photographs Division.

In his published article, Agee cut this and toned other comments down, but he still criticized the orchid trade fairly strongly. Orchid growers were a cartel, he believed, who never really tried to undercut each other even though the agar method meant orchids were being grown cheaply enough to allow them to be sold at much lower prices. It would be possible to make an orchid available to "plenty of people who can't touch it now." Perhaps that larger market didn't exist, but what the growers could be sure of was that "if the orchid became anywhere near as cheap and abundant as the rose, few of its purchasers would be those several sorts and degrees of ladies and gentlemen who at present so satisfactorily enjoy the privilege of supporting it." It was precisely the snob appeal that kept the orchid market going; as Agee bluntly put it, "people like orchids because of their high price."

The artificially high price of orchids was at the heart of Agee's dislike for them; he felt they were "more completely endowed with snob-appeal and with nothing else, than any other commodity I know of," but he admitted that:

I just privately don't like the plain looks of the flower. Any flower is built of course for one special purpose: to propagate itself: of any flower, the private parts and the face are one & the same, and that seems more than all right to me: but it does seem to me that the orchid abuses the privilege.[7]

The claim that orchids "abuse" every flower's privilege of wearing its genitals on its face is a striking one. No doubt the heart of Agee's objection was the spectacle of people spending $12 (about $480/£300 today) on a single orchid at a time when many of their fellow Americans were close to starvation. Given the Depression and Agee's concern for its victims, his reaction is not surprising, but it's hard to imagine any other flower eliciting such a visceral reaction. He really seems to blame the orchids themselves for the reactions they provoke in people. Orchids' sexual and other cultural associations have often brought out strong reactions in people, but Agee seemed to be casting aspersions on the orchid's morals, almost as if he suspected them of manipulating human snobbery to propagate themselves.

Could Agee have been right? Since the 1930s, our understanding of orchids has grown considerably as tissue culture and cloning allowed orchids to be propagated in ever-increasing numbers. The result has been that orchids have been able to colonize an ever-wider variety of ecological niches: airport gift shops, florists, your local supermarket, perhaps your office or your kitchen windowsill. Like the bees and the wasps before us, we have been enlisted into their efforts to win Darwin's game, to leave behind an ever-increasing number of copies of themselves. The symbolic, mythical, religious, and traditional associations of flowers can never be entirely separated from their scientific interpretation; roses, for example, loom so large in the Western imagination that the numbers of species named by scientists has almost certainly been inflated by comparison with flowers that carry less cultural baggage.[8] Perhaps John Boyd's *Pollinators of Eden* isn't really science fiction; it seems that it's not just his imaginary orchids that have "found the ideal animal for their purpose."

If we stop to consider things from the orchid's perspective (which is one of the things I have learned to do by writing this book), what has happened over the past century is that they have found a new pollinator — us. By colonizing our imaginations and modifying our tastes and preferences, we have been persuaded to assist them with their efforts to reproduce and spread (at least, some species and genera have).

Ever since Darwin liberated the plants from their dull, vegetable existence, endowing them with various appetites and the cunning to gratify them, they have loomed ever-larger in our imaginations. Orchids are only one of the kinds of plants that have been reimagined since Darwin in new, and often frightening, ways. Plants have been described as science fiction's "ultimate alien," the things least like ourselves that we can imagine.[9] But perhaps orchids can point us towards a different kind of relationship with plants. In Agee's article, almost the only person associated with the orchid business who seems likeable was the head gardener, a "silent, curiously gentle, all but mystical Swedish ex-sailor," who, Agee says, is "able to say that his orchids 'tell' him 'what they need' and to make that, somehow, not fatuous but almost credible."[10] This gardener sounds somewhat like Darwin, patiently hand-pollinating his orchids, imagining himself as an insect, feeling his way gently into a more intimate relationship to the plants he was trying to understand. The patient gardener recurs regularly in writing about orchids, pictured as a quiet person, in tune with the plants, able to be silent and listen. (As I write this I begin to see why I'm such a poor gardener.) One does not have to be even "semi" mystical (Darwin wasn't) to see that a sensitivity to plants changes our relationship to them, letting us imagine a world where they are no longer simply resources to exploit, things to be put to use, but fellow organisms with their own needs, desires, and goals, all of which we might choose to further instead of ignore. In the *Pollinators of Eden*, Boyd's fictitious botanist, Paul Theaston, tells his lover, Freda Caron, that "as you get to know the orchids, you'll find you're sensitive to them, and to other human beings, in a manner you were never aware of before." Could this bizarre fiction hold an unexpected truth?

Acknowledgments

I would like to thank all those who have helped with the book, particularly Richard Abbott, Sam Alberti, Joseph Arditti, Chad Arment, Richard Bellon, Peter Bernhardt, Daniela Bleichmar, Christophe Bonneuil, Carolyn Burdett, Kath Castillo, Danielle Clode, Justine Cook, Alistair Durie, Matthew Eddy, Florike Egmond, Brent Elliott, Richard England, Martin Evans, Geoffrey Gilbert, Louise Hamilton, Tanja Hammel, Grant Hazlehurst, Oliver Hill-Andrews, Brenton Hobart, Nicholas Jardine, Steve Kershaw, David Knight, Ron Ladouceur, Anne Lemarcis, Jim Livesey, David Livingstone, Geoffrey Lloyd, Roger Luckhurst, George Morgan, Lynne Murphy, Henry Oakeley, Gavin O'Keefe, Charles Nelson, Irene Palmer, Patrick Parrinder, Alison Pearn, Patricia Rain, Alejandra Rojas-Silva, Darrow Schecter, Andre Schuiteman, Julian Shaw, Charlotte Sleigh, Emma Spary, Mike Spencer, Roger Stearn, Joanna Stephens, Phil Stephensen-Payne, Peter Stevens, James Thomson, Charlotte Thurschwell, José Pardo Tomás, Jon Topham, Ted Underwood, Dale Vargas, Nicolas J. Vereecken, Paul White, Susan Whitten, Jesse Willis, and Frank Wu.

This book could never have been written without huge support from the staff of the Royal Botanic Gardens, Kew (especially Craig Brough, Gina Fullerlove, Chris Mills, Virginia Mills, Lynn Parker, Kiri

Ross-Jones, and Tracy Wells). Unpublished letters are reproduced by kind permission of the Board of Trustees of the Royal Botanic Gardens, Kew. Colleagues at the University of Sussex also provided considerable support and encouragement. The increasing numbers of digital resources available to scholars were also invaluable. My particular thanks to: Biodiversity Heritage Library [http://biodiversitylibrary.org/]; The Complete Works of Charles Darwin online [http://darwin-online.org.uk/]; and The Darwin Correspondence Project [http://www.darwinproject.ac.uk/].

Many thanks to Karen Merikangas Darling, Mary Corrado, Kristen Raddatz, Evan White, and all the staff of the University of Chicago Press.

My particular thanks to Phil Cribb and Michael Hutchings, who read the entire manuscript and saved me from some major mistakes (but, naturally, bear no responsibility for any that remain). And to Pam, Max, and Katya, who now all know an orchid when they see one (and have probably seen more than they really wanted to).

Lastly, my biggest thanks go to all the staff of Britain's National Health Service, particularly those of the Sussex Community NHS Trust, without whom I would not be alive today, and to the Macmillan Cancer Support charity (www.macmillan.org.uk), whose nurses have provided me with invaluable support.

Notes

Introduction

1. Susan Orlean, *The Orchid Thief* (London: Vintage, 2000), 47. *Adaptation* (2002) was directed by Spike Jonze, and written by Charlie Kaufman and Donald Kaufman (both played by Nicholas Cage in the film).

2. Patrick Brantlinger, *Rule of Darkness: British Literature and Imperialism, 1830–1914* (Ithaca: Cornell University Press, 1988), 239, 47.

Chapter 1

1. Andrew Dalby, "The Name of the Rose Again; or, What Happened to Theophrastus 'On Aphrodisiacs'?" *Petits Propos Culinaires* 64 (2000): 9–11.

2. Diogenes Laertius, *The Lives and Opinions of Eminent Philosophers*, trans. C. D. Yonge, Bohn's Classical Library (London: Henry G. Bohn, 1853), 195.

3. This translation appears in Dalby, "Theophrastus 'On Aphrodisiacs,'" 11, and earlier in Anthony Preus, "Drugs and Psychic States in Theophrastus' *Historia Plantarum* 9.8–20," in *Theophrastean Studies: On Natural Science, Physics and Metaphysics, Ethics, Religion, and Rhetoric*, ed. William Wall Fortenbaugh and Robert W. Sharples (New Brunswick, NJ: Transaction, 1988), 76; there are no significant differences between the English versions, but Dalby's is slightly less literal and thus more readable.

4. Peter Bernhardt, *Gods and Goddesses in the Garden: Greco-Roman Mythology and the Scientific Names of Plants* (New Brunswick, NJ: Rutgers University Press, 2008), 118.

5. Edward Lee Greene, *Landmarks of Botanical History. A Study of Certain*

Epochs in the Development of the Science of Botany. Part I.— Prior to 1562 A.D., Smithsonian Miscellaneous Collections, Vol. 54 (Washington, DC: Smithsonian Institution Press, 1909), 53.

6. Theophrastus comes from Θειός [*theios*] = divine; and φράσις [*phrasis*] = diction, Laertius, *Lives of Eminent Philosophers*, 195.

7. Ibid., 197–99.

8. The list of these — and many other — books is from ibid., 196–99.

9. Greene, *Landmarks of Botanical History*, 54, 61.

10. Preus, "Drugs and Psychic States in Theophrastus," 88.

11. G. E. R. Lloyd, *Science, Folklore and Ideology: Studies in the Life Sciences in Ancient Greece* (Cambridge: Cambridge University Press, 1983), 1.

12. Jerry Stannard, "Pliny and Roman Botany," *Isis* 56, no. 4 (1965): 420–21.

13. Conway Zirkle, "The Death of Gaius Plinius Secundus (23–79 A.D.)," *Isis* 58, no. 4 (1967): 553–54.

14. Adanson, *Familles des Plantes*, Preface, p. vii, quoted in: Greene, *Landmarks of Botanical History*, 54, 159.

15. Pliny, *Natural History*, XXVI, 62: 189–91.

16. Pliny, *Natural History*, XXVII, 42: 240.

17. Pedianus Dioscorides, *De Materia Medica*, ed. Tess Anne Osbaldeston (Johannesburg, South Africa: Ibidis, 2000), 520–21. Original is from ca. 50–60 CE.

18. Pedanius Dioscorides, *De Materia Medica*, trans. Lily Y. Beck (New York: Olms-Weidmann, 2005), 237–38.

19. Alma Kumbaric, Valentina Savo, and Giulia Caneva, "Orchids in the Roman Culture and Iconography: Evidence for the First Representations in Antiquity," *Journal of Cultural Heritage* 14 (2013).

Chapter 2

1. Martin de la Cruz, *The Badianus Manuscript: (Codex Barberini, Latin 241) Vatican Library; An Aztec Herbal of 1552* (Baltimore: Johns Hopkins University Press, 1940), 314.

2. William Turner, *A New Herball: Parts II and III*, Vol. 2 (Cambridge: Cambridge University Press, 1568 [1992]), 586.

3. "Of Dogs Stones: The Vertues," John Gerard, *The Herball: Or, Generall Historie of Plants* [photoreprint of the 1597 ed. published by J. Norton, London], Vol. 1, chapter 98, The English Experience, Its Record in Early Printed Books Published in Facsimile (Norwood, NJ: W. J. Johnson, 1974), 158.

4. Elizabeth L Eisenstein, *The Printing Revolution in Early Modern Europe* (Cambridge: Cambridge University Press, 1983), 12–40.

5. Paula Findlen, "Natural History," in *The Cambridge History of Science*, Vol. 3, *Early Modern Science*, ed. Katharine Park and Lorraine Daston (Cambridge: Cambridge University Press, 2006), 439–40.

6. J. B. Trapp, "Dioscorides in Utopia," *Journal of the Warburg and Courtauld Institutes* 65 (2002): 259–60.

7. Findlen, "Natural History," 440–41.

8. Logbook, October 19 and 21, 1492, quoted in: ibid., 448.

9. Simon Varey and Rafael Chabrán, "Medical Natural History in the Renaissance: The Strange Case of Francisco Hernández," *Huntington Library Quarterly* 57, no. 2 (1994): 133; Findlen, "Natural History," 451–52.

10. The people whom we refer to as the Aztecs were only one of many groups living in Central America who spoke Náhuatl, whom modern scholars refer to collectively as Náhuas. The Aztecs called themselves "Mexicah," but neither de la Cruz nor Badiano were Mexicah, so the term "Náhua" is most appropriate. Millie Gimmel, "Reading Medicine in the *Codex de la Cruz Badiano*," *Journal of the History of Ideas* 69, no. 2 (2008): 177.

11. Ibid., 175–77.

12. Ralph Bauer, "A New World of Secrets: Occult Philosophy and Local Knowledge in the Sixteenth-Century Atlantic," in *Science and Empire in the Atlantic World*, ed. James Delbourgo and Nicholas Dew (London: Routledge, 2008), 99–101.

13. Henry Lowood, "The New World and the European Catalog of Nature," in *America in European Consciousness, 1493–1750*, ed. Karen Ordahl Kupperman (Chapel Hill: University of North Carolina Press, 1995), 303–5; Bauer, "New World of Secrets," 110.

14. Findlen, "Natural History," 444–55.

15. Donovan S. Correll, "Vanilla: Its Botany, History, Cultivation and Economic Import," *Economic Botany* 7, no. 4 (1953): 295.

16. Richard H. Drayton, *Nature's Government: Science, Imperial Britain and the "Improvement" of the World* (New Haven: Yale University Press, 2000), 13–14.

17. Eisenstein, *Printing Revolution*, 51.

18. Agnes Arber, *Herbals: Their Origin and Evolution. A Chapter in the History of Botany 1470–1670*, 3rd ed. (Cambridge: Cambridge University Press, 1990), 58–59; Thomas Archibald Sprague, "The Evolution of Botanical Taxonomy from Theophrastus to Linnaeus," in *Linnean Society's Lectures on the Development of Taxonomy* (London: Linnean Society of London, 1948–49), 5.

19. Tragus, *De Stirpium Historia*, Vol. 2, chapter 82 (Zurich, 1552), 784. (This is the Latin translation of the *Kreütter Buch*.) Greene, *Landmarks of Botanical History*, 1024n32.

20. Arber, *Herbals*, 248–49; Allen G. Debus, *The Chemical Philosophy: Paracelsian Science and Medicine in the Sixteenth and Seventeenth Centuries* (New York: Science History, 1977), 46–48.

21. Arber, *Herbals*, 250–51.

22. Valeria Finucci, *The Manly Masquerade: Masculinity, Paternity and Castration in the Italian Renaissance* (Durham, NC: Duke University Press, 2003), 18.

23. John Baptista Porta, *Natural Magick* (London: Thomas Young and Samuel Speed, 1658). Quoted in: Debus, *Chemical Philosophy*, 34.

24. Finucci, *Manly Masquerade*, 18.

25. Debus, *Chemical Philosophy*, 117–23.

26. Oswald Croll and Johann Hartmann, *A Treatise of Oswaldus Crollius of Signatures of Internal Things; or, A True and Lively Anatomy of the Greater and Lesser World* (London: Starkey, 1669), 1–6. Lobelius was the Latinized name of Mathias de l'Obel.

27. Arber, *Herbals*, 252–54.

28. John Parkinson, *Paradisi in Sole Paradisus Terrestris, or, a Garden of All Sorts of Pleasant Flowers Which Our English Ayre Will Permitt to Be Noursed Vp: With a Kitchen Garden of All Manner of Herbes, Rootes, & Fruites, for Meate or Sause Vsed with Vs, and an Orchard of All Sorte of Fruitbearing Trees and Shrubbes Fit for Our Land Together with the Right Orderinge Planting & Preseruing of Them and Their Vses & Vertues* (London: Humfrey Lownes and Robert Young, 1629), 67.

29. Steven Shapin, *The Scientific Revolution* (Chicago: University of Chicago Press, 1996).

30. John Parkinson, J. Mackie, and University of Cambridge Dept. of Plant Sciences, *Theatrum Botanicum: The Theater of Plants: Or, an Herball of a Large Extent; Containing Therein a More Ample and Exact History and Declaration of the Physicall Herbs and Plants That Are in Other Authours, Encreased by the Accesse of Many Hundreds of New, Rare, and Strange Plants from All the Parts of the World, with Sundry Gummes, and Other Physicall Materials, Than Hath Beene Hitherto Published by Any before; and a Most Large Demonstration of Their Natures and Vertues; Shewing Withall the Many Errors, Differences, and Oversights of Sundry Authors That Have Formerly Written of Them; and a Certaine Confidence, or Most Probable Conjecture of the True and Genuine Herbes and Plants. Distributed into Sundry Classes or Tribes, for the More Easie Knowledge of the Many Herbes of One Nature and Property, with the Chiefe Notes of Dr. Lobel, Dr. Bonham, and Others Inserted Therein* (London: Tho. Cotes, 1640), 1346.

31. *Stirpium Historiae Pemptades Sex* (1581), quoted in: Arber, *Herbals*, 255–56.

32. Greene, *Landmarks of Botanical History*, 54, 237.

33. Julius Rembert Dodoens, *Stirpium Historiae Pemptades Sex* (Antwerp: Christopher Plantin, 1583); Greene, *Landmarks of Botanical History*, 1, 858–59.

34. Mark A. Waddell, "Magic and Artifice in the Collection of Athanasius Kircher," *Endeavour* 34, no. 1 (2009), 30–34.

35. Jakob Breyne, *Exoticarum Aliarumque Minus Cognitarum Plantarum* (1678). Quoted in: Oakes Ames, *Orchids in Retrospect: A Collection of Essays on the Orchidaceae* (Cambridge, MA: Botanical Museum of Harvard University, 1948), 1.

Chapter 3

1. William T. Stearn, "Two Thousand Years of Orchidology," in *Proceedings of the Third World Orchid Conference* (London: Royal Horticultural Society, 1960), 32.

2. Drayton, *Nature's Government*, 14–15. See also: Bruno Latour, *Science in Action: How to Follow Scientists and Engineers through Society* (Cambridge, MA: Harvard University Press, [1987] 1994), 219–25.

3. Julius Sachs, *History of Botany, 1530–1860*, trans. Henry E. F. Garnsey (Oxford: Oxford University Press, 1906), 30.

4. *Stirpium Adversaria Nova* (1576), quoted in: ibid., 31–32.

5. Charles Schweinfurth, "Classification of Orchids," in *The Orchids: A Scientific Survey*, ed. Carl Leslie Withner (New York: Ronald Press, 1959), 16.

6. Lowood, "European Catalog," 295; Drayton, *Nature's Government*, 15–16; Arthur James Cain, "Rank and Sequence in Caspar Bauhin's Pinax," *Botanical Journal of the Linnean Society* 114 (1994): 311–56.

7. Drayton, *Nature's Government*, 17.

8. Ibid., 17–18.

9. Hans Sloane, *A Voyage to the Islands Madera, Barbados, Nieves, S. Christophers and Jamaica: With the Natural History of the Herbs and Trees, Four-Footed Beasts, Fishes, Birds, Insects, Reptiles, &C. Of the Last of Those Islands; to Which Is Prefix'd, an Introduction, Wherein Is an Account of the Inhabitants, Air, Waters, Diseases, Trade, &C. Of That Place, with Some Relations Concerning the Neighbouring Continent, and Islands of America. Illustrated with Figures of the Things Described, Which Have Not Been Heretofore Engraved* (London: B. M. for the Author, 1707–1725), preface.

10. Varey and Chabrán, "Strange Case of Francisco Hernández," 141.

11. Peter Dear, *Revolutionizing the Sciences: European Knowledge and Its Ambitions, 1500–1700* (Basingstoke: Palgrave, 2001), 111–13.

12. Jorge Cañizares-Esguerra, "Iberian Science in the Renaissance: Ignored How Much Longer?" *Perspectives on Science* 12, no. 1 (2004): 86–124.

13. Lowood, "European Catalog," 295–315.

14. Sachs, *History of Botany*, 76–77.

15. Stearn, "Two Thousand Years of Orchidology," 32; Charlie Jarvis and Phillip Cribb, "Linnaean Sources and Concepts of Orchids," *Annals of Botany* 104 (2009): 367.

16. Van Rheede, preface to Vol. 3 of *Hortus Malabaricus*. Quoted in: H. Y. Mohan Ram, "On the English Edition of Van Rheede's *Hortus Malabaricus* by K. S. Manilal (2003)," *Current Science* 89, no. 10 (2005): 1673–74.

17. Stearn, "Two Thousand Years of Orchidology," 33.

18. Jarvis and Cribb, "Linnaean Concepts of Orchids," 367.

19. Wilfrid Blunt, *The Compleat Naturalist: A Life of Linnaeus* (London: Collins, 1971).

20. Lisbet Koerner, "Carl Linnaeus in His Time and Place," in *Cultures of Natural History*, ed. Nicholas Jardine, James A. Secord, and Emma Spary (Cambridge: Cambridge University Press, 1996), 148–49.

21. Peter Harrison, "Linnaeus as a Second Adam? Taxonomy and the Religious Vocation," *Zygon* 44, no. 4 (2009): 879–80.

22. Londa Schiebinger, "The Private Life of Plants: Sexual Politics in Carl Linnaeus and Erasmus Darwin," in *Science and Sensibility: Gender and Scientific Enquiry, 1780–1945*, ed. Marina Benjamin (Oxford: Blackwell, 1991), 122–23.

23. Carl Linnaeus, *Systema Naturae, sive Regna Tria Naturae Systematice Proposita per*

Classes, Ordines, Genera, & Species (Leiden, 1735). Quoted in: Staffan Müller-Wille, "Systems and How Linnaeus Looked at Them in Retrospect," *Annals of Science* 70, no. 3 (2013): 311.

24. Ibid., 314.

25. Linnaeus, *Philosophia Botanica* (1751), aphorism 77. Quoted in: Staffan Müller-Wille, "Collection and Collation: Theory and Practice of Linnaean Botany," *Studies in History and Philosophy of Biological and Biomedical Sciences* 38 (2007): 553.

26. Lisbet Koerner, *Linnaeus: Nature and Nation* (Cambridge, MA: Harvard University Press, 1999), 114.

27. Carl Linnaeus, *Critica botanica* (1737). Quoted in: Müller-Wille, "Collection and Collation," 560.

28. Carl Linnaeus, *Genera Plantarum Eorumque Characters Naturales Secundum Numerum, Figuram, Situm, Proportionem Omnium Fructificationis Partium, sixth edition* (Stockholm, 1764), "Ordines naturales," [aph. 10]. Quoted in: Staffan Müller-Wille, "Systems and How Linnaeus Looked at Them in Retrospect," *Annals of Science* 70, no. 3 (2013): 305–17.

29. Carl Linnaeus, *Philosophia Botanica in qua Explicantur Fundamenta Botanica cum Definitionibus Partium, Exemplis Terminorum, Observationibus Rariorum* (Stockholm, 1751). Quoted in: Müller-Wille, "Systems and How Linnaeus Looked at Them," 314.

30. Jarvis and Cribb, "Linnaean Concepts of Orchids," 367–71.

31. James Larson, "Linnaeus and the Natural Method," *Isis* 58, no. 3 (1967): 304–20; Peter F. Stevens, *The Development of Biological Systematics: Antoine-Laurent de Jussieu, Nature, and the Natural System* (New York: Columbia University Press, 1994).

32. Linnaeus, *Præludia Sponsaliorum Plantarum* (On the prelude to the wedding of plants), XVI, p. 14. Quoted in: Larson, "Linnaeus and the Natural Method," 306.

33. Schiebinger, "Private Life of Plants."

34. Janet Browne, "Botany for Gentlemen: Erasmus Darwin and *The Loves of the Plants*," *Isis* 80, no. 304 (1989): 600.

35. Díaz del Castillo Bernal, *True History of the Conquest of New Spain (Historia Verdadera de la Conquista de la Nueva España)*. Quoted in: Sophie D. Coe and Michael D. Coe, *The True History of Chocolate* (London: Thames & Hudson, 1996 [2013]).

36. Hernández, *Obras Completas*. Quoted in: Coe and Coe, *True History of Chocolate*.

37. Cruz, *Badianus Manuscript*, 314; Johannes de Laet, "Manuscript Translation of Hernández," trans. Rafael Chabrán, Cynthia L. Chamberlin, and Simon Varey, in *The Mexican Treasury: The Writings of Dr. Francisco Hernández*, ed. Simon Varey (Stanford: Stanford University Press, 2000), 167.

38. The story appears in Patricia Rain, *Vanilla: The Cultural History of the World's Favorite Flavor and Fragrance* (Los Angeles: Jeremy P. Tarcher, 2004), where Zimmermann's first name is mistakenly given as "Bezaar."

39. Coe and Coe, *True History of Chocolate*; Roy Porter and Mikuláš Teich, eds., *Drugs and Narcotics in History* (Cambridge: Cambridge University Press, 1995), 29–30.

40. Louis Liger, *The Compleat Florist: Or, the Universal Culture of Flowers, Trees and*

Shrubs; Proper to Imbellish Gardens (London: Printed for Benj. Tooke, at the Temple-Gate, Fleetstreet, 1706), 269.

41. Annette Giesecke, *The Mythology of Plants: Botanical Lore from Ancient Greece and Rome* (Los Angeles: J. Paul Getty Museum, 2014), 23.

42. Bernhardt, *Gods and Goddesses*, 59; Giesecke, *Mythology of Plants*, 30.

43. Giesecke, *Mythology of Plants*, 49.

44. Liger, *Compleat Florist*, 270.

45. Renate Blumenfeld-Kosinski, "The Scandal of Pasiphae: Narration and Interpretation in the 'Ovide Moralisé,'" *Modern Philology* 93, no. 3 (1996): 307–26.

46. My thanks to Emma Spary for this suggestion and for much useful information on Liger's context.

47. Liger, *Le Voyageur Fidèle, ou le Guide des Etrangers dans la Ville de Paris, qui Enseigne Tout ce qu'il y a de Plus Curieux à Voir*, 1715: 356). Quoted in: E. C. Spary, *Eating the Enlightenment: Food and the Sciences in Paris* (Chicago: University of Chicago Press, 2012), 115.

48. Spary, *Eating the Enlightenment*, 121. See also: Geoffrey V. Sutton, *Science for a Polite Society* (Boulder, CO: Westview, 1995), 8.

49. Sutton, *Science for a Polite Society*, 8–15.

50. Grace Greylock Niles, *Bog-Trotting for Orchids* (New York: G. P. Putnam's Sons, 1904), 109–10.

51. Even scholarly works sometimes include the myth without explicitly stating that it has no classical sources, e.g., Bernhardt, *Gods and Goddesses*.

Chapter 4

1. Relative prices have been calculated relative to average wages by using the "labour value" calculation at MeasuringWorth.com. Conversions to US$ were done using the average 2014 exchange rate.

2. Harry James Veitch, "A Retrospect of Orchid Culture," in *A Manual of Orchidaceous Plants: Cultivated under Glass in Great Britain* (London: James Veitch and Sons, 1887–94), 109.

3. Stearn, "Two Thousand Years of Orchidology," 35–36.

4. Sander to Roezl, c.1883. Quoted in: Arthur Swinson, *Frederick Sander: The Orchid King* (London: Hodder and Stoughton, 1970), 29. See also: Merle A. Reinikka, *A History of the Orchid*, 2nd ed. (Portland, OR: Timber, 1995), 61–64.

5. James Brooke, *The Fairfield Orchids, a Descriptive Catalogue of the Species and Varieties Grown by James Brooke and Co., Fairfield Nurseries* (Manchester: James Brooke, 1872), 7.

6. Quoted in: Veitch, "Retrospect of Orchid Culture," 110.

7. Peter Bernhardt, *Wily Violets and Underground Orchids: Revelations of a Botanist* (New York: Vintage, 1990), 186–87.

8. These are described in the *Botanical Register*, Vol. 3, 1817 (following plate 220, unpaginated).

9. Veitch, "Retrospect of Orchid Culture," 113.

10. Ibid., 114–16.

11. Quoted in Phillip Cribb, Mike Tibbs, John Day, and Kew Royal Botanic Gardens, *A Very Victorian Passion: The Orchid Paintings of John Day, 1863 to 1888* (Kew: Royal Botanic Gardens, 2004), 11; and Veitch, "Retrospect of Orchid Culture," 116.

12. James Bateman, *The Orchidaceae of Mexico and Guatemala* (London: For the author, J. Ridgway & Sons, 1837–1843), 10–14.

13. Ibid., 3–5.

14. E. Charles Nelson, *John Lyons and His Orchid Manual* (Kilkenny, Ireland: Boethius, 1983), 20–22, 46–52.

15. John Charles Lyons, *Remarks on the Management of Orchidaceous Plants, with a Catalogue of Those in the Collection of J. C. Lyons, Ladiston. Alphabetically Arranged, with Their Native Countries and a Short Account of the Mode of Cultivation Adopted* (Ladiston: Self-published, 1843), n.p., 96.

16. In works such as *Sertum Orchidaceum* (1838) and *The Genera and Species of Orchidaceous Plants* (1830–40). See: Schweinfurth, "Classification of Orchids," 20–23; Brent Elliott, "The Royal Horticultural Society and Its Orchids: A Social History," *Occasional Papers from the RHS Lindley Library* 2 (2010): 8–11.

17. Lyons, *Management of Orchidaceous Plants*; Emphasis in original.

18. Ibid., 10–15.

19. Ibid., 4.

20. Jeffrey A. Auerbach, *The Great Exhibition of 1851: A Nation on Display* (New Haven: Yale University Press, 1999); James Buzard, Joseph W. Childers, and Eileen Gillooly, eds., *Victorian Prism: Refractions of the Crystal Palace* (Charlottesville: University of Virginia Press, 2007); Isobel Armstrong, *Victorian Glassworlds: Glass Culture and the Imagination 1830–1880* (Oxford: Oxford University Press, 2008).

21. Ker, introduction to Williams' "Orchids for the Million," *Gardeners' Chronicle*, no. 20 (May 17, 1851): 308.

22. Henry Williams, "[Obituary] Bernard S. Williams," *Orchid Album* 9 (1891).

Chapter 5

1. This previously obscure episode in Dutch history became well-known in Victorian Britain following the publication of a curious book called *Memoirs of Extraordinary Popular Delusions and the Madness of Crowds*, by Charles Mackay (1841).

2. [John Roby Leifchild], "Review of Darwin, CR 'on the Various Contrivances by Which British and Foreign Orchids Are Fertilized by Insects,'" *Athenaeum*, no. 1804 (1862): 683.

3. Anonymous, "Mr. Darwin's Orchids," *Saturday Review of Politics, Literature, Science, and Art* (1862): 486.

4. J. D. Hooker to C. R. Darwin, January 29, 1844 (emphasis added). Darwin Correspondence Database, http://www.darwinproject.ac.uk/entry-734.

5. Jim Endersby, *Imperial Nature: Joseph Hooker and the Practices of Victorian Science* (Chicago: University of Chicago Press, 2008).

6. Francis Darwin, *The Life and Letters of Charles Darwin*, rev. ed., Vol. 3 (London: John Murray, 1888), 263. See also: Retha Edens-Meier and Peter Bernhardt, eds., *Darwin's Orchids: Then and Now* (Chicago: University of Chicago Press, 2014), 6.

7. Mea Allan, *Darwin and His Flowers: The Key to Natural Selection* (London: Faber and Faber, 1977), 195.

8. Charles Darwin, *On the Various Contrivances by Which British and Foreign Orchids Are Fertilised by Insects, and on the Good Effects of Intercrossing* (London: John Murray, 1862), 1–2.

9. Anonymous, "Darwin's Orchids."

10. Richard Bellon, "Inspiration in the Harness of Daily Labor: Darwin, Botany, and the Triumph of Evolution, 1859–1868," *Isis* 102 (2011), 393–420.

11. William Bernhard Tegetmeier, "Darwin on Orchids," *Weldon's Register of Facts and Occurrences* (1862).

12. Anonymous, "Fertilization of Orchids. By Charles Darwin," *Parthenon* 1, no. 6 (1862): 177.

13. Tegetmeier, "Darwin on Orchids."

14. John Murray to Charles Darwin, July 29 [1874], Frederick Burkhardt, James A. Secord, and the Editors of the Darwin Correspondence Project, eds., *The Correspondence of Charles Darwin, Vol. 24, 1874* (Cambridge: Cambridge University Press, 2015), 394.

15. Joseph Dalton Hooker, "[Review of] Darwin, C. R. On the Various Contrivances by Which British and Foreign Orchids Are Fertilized by Insects," *Natural History Review* 2, no. 8 (1862): 371.

16. Darwin, *Orchids*, 2.

17. Although the two terms are sometimes used interchangeably, pollination and fertilization are not the same thing. Pollination is the depositing of pollen on the receptive surface, but the pollen grains then grow pollen tubes that carry the male gametes down the style to the ovary, where fertilization occurs (i.e., the male and female gametes fuse to form a zygote).

18. Charles Darwin, *On the Origin of Species by Means of Natural Selection: Or the Preservation of Favoured Races in the Struggle for Life* (London: John Murray, 1859), 62–63.

19. Jim Endersby, "Darwin on Generation, Pangenesis and Sexual Selection," in *Cambridge Companion to Darwin*, ed. M. J. S. Hodge and G. Radick (Cambridge: Cambridge University Press, 2003), 69–91.

20. Darwin, *Orchids*, 4.

21. Ibid., 15.

22. Ibid., 1.

23. W. Paley, *Natural Theology: Or, Evidences of the Existence and Attributes of the Deity*, 12th ed. (London: J. Faulder, 1809), 11.

24. Much of my thinking on these topics was originally inspired by "Beautiful Contrivance: Science, Religion and Language in Darwin's Fertilization of Orchids," by Richard England, Salisbury University, an unpublished copy of a seminar paper given in October

2001. For natural theology generally, see: John Hedley Brooke, *Science and Religion: Some Historical Perspectives* (Cambridge: Cambridge University Press, 1991); Thomas Dixon, *Science and Religion: A Very Short Introduction* (Oxford: Oxford University Press, 2008).

25. Anonymous, "Fertilization of Orchids," 178.

26. Tegetmeier, "Darwin on Orchids."

27. Anonymous, "Darwin's Orchids."

28. George (8th Duke of Argyll) Campbell, "The Supernatural [Review of Darwin on Orchids and Other Works]," *Edinburgh Review* 116, no. 236 (1862): 392. As an example, Campbell cites Darwin's comment that in one orchid: "we have the rostellum partially closing the mouth of the nectary, like a trap placed in a run for game," Darwin, *Orchids*, 30.

29. Paley, *Natural Theology*, 2–12.

30. C. Darwin to Asa Gray, November 23 [1862]. Darwin Correspondence Database, http://www.darwinproject.ac.uk/entry-3820

31. For Gray's religious views, see A. Hunter Dupree, *Asa Gray: American Botanist, Friend of Darwin*, 1st paperback ed. (Baltimore: Johns Hopkins University Press, 1959 [1988]), 380–81; and http://www.darwinproject.ac.uk/religion-historical-resources

32. Asa Gray, "[Review of C. R. Darwin] On the Various Contrivances by Which British and Foreign Orchids Are Fertilised by Insects," *American Journal of Science and Arts* New Series 34 (1862): 139.

33. George (8th Duke of Argyll) Campbell, "What Is Science?" *Good Words* 26 (1885): 243–44.

34. C. Darwin to Asa Gray, May 22 [1860]: Frederick Burkhardt et al., eds., *The Correspondence of Charles Darwin. Vol. 8, 1860* (Cambridge: Cambridge University Press, 1993), 224.

35. Bernard Lightman and Gowan Dawson, eds., *Victorian Scientific Naturalism: Community, Identity, Continuity* (Chicago: University of Chicago Press, 2014).

36. Gray, "[Darwin] Orchids," 139.

37. Asa Gray to Darwin, July 2, 1862. Darwin Correspondence Database, http://www.darwinproject.ac.uk/entry-3637

38. Darwin to Asa Gray, July 23[–4], [1862], Darwin Correspondence Database, http://www.darwinproject.ac.uk/entry-3662

39. [Leifchild], "Review of Orchids," 684.

40. Darwin, *Orchids*, 210.

41. *The Various Contrivances by Which British and Foreign Orchids Are Fertilised by Insects, and on the Good Effects of Intercrossing*, 2nd ed. (London: John Murray, 1877 [1904]), 175–76.

42. Darwin quoted this from Lindley's *The Vegetable Kingdom* (1853, p. 178), in *Orchids*, 236.

43. Edens-Meier and Bernhardt, *Darwin's Orchids: Then and Now*, 12.

44. Darwin, *Orchids*, 244. The "distinguished botanist" was Alphonse de Candolle. Michael T. Ghiselin, "Darwin: A Reader's Guide," *Occasional Papers of the California Academy of Sciences* 155 (2009): 24.

45. Darwin, *Orchids*, 306.

46. Asa Gray, ed., *Darwiniana: Essays and Reviews Pertaining to Darwinism*, ed. A. Hunter Dupree (Cambridge: Harvard University Press, 1963).

47. John Beatty, "Chance Variation: Darwin on Orchids," *Philosophy of Science* 73, no. 5 (2006), 629–41. See also: James G. Lennox, "Darwin *Was* a Teleologist," *Biology and Philosophy* 8, no. 4 (1993), 409–21.

48. Asa Gray to Darwin, July 2, 1862. Darwin Correspondence Database, http://www.darwinproject.ac.uk/entry-3637

49. Asa Gray to Darwin, July 7, 1863. Darwin Correspondence Database, http://www.darwinproject.ac.uk/entry-4234

50. Darwin to Asa Gray, Aug 4 [1863], Darwin Correspondence Database, http://www.darwinproject.ac.uk/entry-4262

51. Darwin, *Orchids*, 130–33, 349–50.

52. Ibid., 198.

53. Joseph Arditti et al., "'Good Heavens What Insect Can Suck It'–Charles Darwin, *Angraecum sesquipedale* and *Xanthopan morganii praedicta*," *Botanical Journal of the Linnean Society* 169 (2012): 408–9.

54. Darwin, *Orchids*, 198.

55. C. R. Darwin to J. D. Hooker, January 25 [1862], Darwin Correspondence Database, http://www.darwinproject.ac.uk/entry-3411

56. C. R. Darwin to J. D. Hooker, January 30 [1862]. Darwin Correspondence Database, http://www.darwinproject.ac.uk/entry-3421

57. Darwin, *Orchids*, 198.

58. Ibid., 201–3.

59. Arditti et al., "What Insect Can Suck It," 419–25.

60. Bernhardt, *Wily Violets*, 199–200.

61. St. G. J. Mivart to C. R. Darwin, June 11, 1870; and C. R. Darwin to St. G. J. Mivart, June 13 [1870]. Darwin Correspondence Database, http://www.darwinproject.ac.uk/entry-7227 and entry-7228a

62. St. George Jackson Mivart and Making of America Project, *On the Genesis of Species by St. George Mivart, F. R. S.* (New York: D. Appleton, 1871), 67–68.

63. Darwin, *Orchids*, 68.

64. Darwin, C. "Recollections of the Development of My Mind and Character," published as: *Autobiographies*, ed. Michael Neve and Sharon Messenger (Harmondsworth: Penguin Books, 1887/1903 [2003]).

Chapter 6

1. Darwin to Hooker, Jan 13 [1863]. Darwin Correspondence Database, http://www.darwinproject.ac.uk/entry-3913

2. Darwin to Hooker, [July 6, 1861]. Darwin Correspondence Database, http://www.darwinproject.ac.uk/entry-3200

3. Darwin, *Orchids*, 158.

4. Brooke, *Fairfield Orchids*, 2.

5. Swinson, *Frederick Sander*.

6. Frederick Boyle, *The Woodlands Orchids Described and Illustrated, with Stories of Orchid-Collecting* (London: Macmillan, 1901), 17–20.

7. Alice M. Coats, *The Quest for Plants: A History of Horticultural Explorers* (London: Studio Vista, 1969), 348–50.

8. Swinson, *Frederick Sander*, 23–24, 7.

9. Ibid., 29.

10. Stearn, "Two Thousand Years of Orchidology," 35–36; Lyons, *Management of Orchidaceous Plants*, 7.

11. Swinson, *Frederick Sander*, 29–30.

12. Percy Collins, "The Romance of Auctioneering," *Strand Magazine* 31, no. 181 (1906): 100–106. The precise date of the sale is not given.

13. W. A. Stiles, "Orchids," *Scribner's Monthly* 15, no. 2 (1894): 191.

14. C. G. G. J. van Steenis and M. J. van Steenis-Kruseman, *Flora Malesiana*, Series I, Spermatophyta, ed. C. G. G. J. van Steenis, Vol. 1 (Djakarta: Noordhoff-Kolff, 1950), 360.

15. Swinson, *Frederick Sander*, 81–83.

16. Stiles, "Orchids," 192.

17. Swinson, *Frederick Sander*, 84–85, 103.

18. Micholitz to Sander, Macassar June 2, 1891. Frederick Sander Papers (Archives, Royal Botanic Gardens, Kew), Vol. 11, Letters from Wilhelm Micholitz.

19. Stiles, "Orchids," 193–94; Michael Sidney Tyler-Whittle, *The Plant Hunters, Being an Examination of Collecting, with an Account of the Careers & the Methods of a Number of Those Who Have Searched the World for Wild Plants* (Philadelphia: Chilton, 1970), 9–10.

20. "The Great Orchid Sale," *Standard*, October 17, 1891.

21. The story seems to originate with Boyle, who ascribes it to Paxton but gives no source. Frederick Boyle, *About Orchids: A Chat* (London: Chapman & Hall, 1893), 173–74.

22. John Lindley, *Collectanea Botanica, or Figures and Botanical Illustrations of Rare and Curious Exotic Plants* (London: Richard and Arthur Taylor, Shoe-Lane, 1821), t.33.

23. Boyle, *About Orchids*, 177–79.

24. Tyler-Whittle, *The Plant Hunters*, 124.

25. Robert Allen Rolfe, "*Cattleya labiata* and Its Habitat," *Orchid Review* 8, no. 96 (1900): 365. The myth of Swainson using the orchids as packaging material is still repeated in generally reliable books, e.g., Reinikka, *History of the Orchid*, 23.

26. *Gardeners' Chronicle*, Third Series 10, no. 248 (September 26, 1891): 363.

27. *Gardeners' Chronicle*, Third Series 10, no 150 (October 10, 1891): 414.

28. "Great Orchid Sale."

29. Micholitz to Sander, Macassar February 10, 1891. Frederick Sander Papers (Archives, Royal Botanic Gardens, Kew), Vol. 11, Letters from Wilhelm Micholitz.

30. Micholitz to Sander, Macassar June 2, 1891. Frederick Sander Papers (Archives, Royal Botanic Gardens, Kew), Vol. 11, Letters from Wilhelm Micholitz. Extracts were published in: Swinson, *Frederick Sander*, 106.

31. Swinson, *Frederick Sander*, 106–7.

32. "An Orchid Sale," in: Boyle, *About Orchids*. The book is a collection of newspaper articles; Boyle doesn't give the original publication details but his article must have first appeared in print sometime between October 1891 and the collection's appearance in 1893.

33. Boyle, "The Story of an Orchid," *Ludgate* 4 (1897): 508.

34. Henry Ogg Forbes, "Three Months' Exploration in the Tenimber Islands, or Timor Laut," *Proceedings of the Royal Geographical Society and Monthly Record of Geography* New Monthly Series 6, no. 3 (1884): 113–29; William Turner Thiselton-Dyer and Daniel Oliver, "Report of the Botany of Mr. H. O. Forbes's Expedition to Timor-Laut," *Journal of the Linnean Society (Botany)* 21 (1885): 370–374; Joseph Dalton Hooker, "Dendrobium phalae-nopsis," *Curtis's Botanical Magazine, Comprising the Plants of the Royal Gardens of Kew and of Other Botanical Establishments in Great Britain* Third Series 41 (1885).

35. Swinson, *Frederick Sander*, 126, 42–44.

36. Frederick Boyle and Joseph Godseff, *The Culture of Greenhouse Orchids: Old System and New* (London: Chapman & Hall, 1902).

37. Boyle, *About Orchids*, 32.

38. Lady Charlotte, "Most Rare: Flowers That Cost Lives to Secure," *Daily Mail*, no. 3 (1896): 7.

39. Stiles, "Orchids," 193–94.

40. "M. Hamelin's Adventures in Madagascar," *Standard*, July 25, 1893.

41. "Eulophiella elisabethae," *Gardeners' Chronicle* Third Series 14, no. 340 (July 1, 1893): 14.

42. Sander and Co., "Great Orchid Sale" [advertisement], *Gardeners' Chronicle* Third Series 14, no. 340 (July 1, 1893): 7.

43. *Nottinghamshire Guardian* (London, England), July 8, 1893, 5. The *Liverpool Echo*'s piece also appeared on July 8, 1893.

44. *Yorkshire Herald*, and *York Herald* (York, England), July 10, 1893, 3.

45. "London Letter," *Garden and Forest* 6, no. 281 (1893): 294.

46. *Standard* (London, England), October 25, 1893, 3.

47. "Madagascar:—Eulophiella elisabethae," *Gardeners' Chronicle* Third Series 14, no. 358 (1893): 556.

48. "The New Orchid, 'Eulophiella elisabethæ,' of Madagascar," *Standard* (London, England), October 27, 1893, 3. The species had been described and named by Robert Allen Rolfe in *Lindenia* Part 16, 5 (1892): 29–30.

49. Lucien Linden, "History of the Introduction of *Eulophiella elisabethae*," *Lindenia* Part 16, 3 (1892): 29–30.

50. "Orchid Hunting: A Dangerous Pastime," *Singapore Free Press and Mercantile Advertiser*, October 18, 1901, 3.

51. Boyle, *Woodlands Orchids*, 117.

52. Brooke, *Fairfield Orchids*, 10.

53. Gardens full of exotic plants had been seen as trophies of imperial conquest since ancient Roman times. Giesecke, *Mythology of Plants*, 18–19.

54. Stiles, "Orchids," 193.

55. Ibid., 196.

56. Boyle, *About Orchids*, 26.

Chapter 7

1. Benjamin Samuel Williams, *The Orchid-Grower's Manual: Containing Brief Descriptions of Upwards of Four Hundred Species and Varieties of Orchidaceous Plants: Together with Notices of Their Times of Flowering and Most Approved Modes of Treatment: Also, Plain and Practical Instructions Relating to the General Culture of Orchids: And Remarks on the Heat, Moisture, Soil, and Seasons of Growth and Rest Best Suited to the Several Species. Seventh Edition, Enlarged and Revised to the Present Time by Henry Williams* (London Victoria and Paradise Nurseries, 1894), 17.

2. Samuel Moskowitz, ed., *Science Fiction by Gaslight: A History and Anthology of Science Fiction in the Popular Magazines, 1891–1911*, 2nd ed. (Westport, CT: Hyperion, 1974), 25.

3. Charlotte F. Otten, "Ophelia's 'Long Purples' or 'Dead Men's Fingers,'" *Shakespeare Quarterly* 30, no. 3 (1979): 397–402.

4. Quoted in: Londa Schiebinger, "Gender and Natural History," in *Cultures of Natural History*, ed. Nicholas Jardine, James A. Secord, and Emma Spary (Cambridge: Cambridge University Press, 1996), 163–77.

5. Schiebinger, "Private Life of Plants," 128.

6. Brooke, *Fairfield Orchids*, 10.

7. For this, and several of the other stories I discuss, I am deeply indebted to the work of Chad Arment, who has spent years tracking down these stories and has published three weird and wonderful collections of them: Chad Arment, ed., *Botanica Delira: More Stories of Strange, Undiscovered, and Murderous Vegetation* (Landisville, PA: Coachwhip, 2010); *Flora Curiosa: Cryptobotany, Mysterious Fungi, Sentient Trees, and Deadly Plants in Classic Science Fiction and Fantasy*, 2nd (rev.) ed. (Greenville, OH: Coachwhip, 2013); and *Arboris Mysterius: Stories of the Uncanny and Undescribed from the Botanical Kingdom* (Greenville, OH: Coachwhip, 2014). Some of these stories, and others, can also be found in: Carlos Cassaba, ed. *Roots of Evil: Beyond the Secret Life of Plants* (London: Corgi Books, 1976).

8. Hawthorne, "Rappaccini's Daughter" (1844), in: Arment, *Flora Curiosa*, 7–37.

9. Fred M. White, "The Purple Terror," *Strand Magazine* 18, no. 105 (1896): 244.

10. Charles Darwin and Francis Darwin, *The Power of Movement in Plants* (London: John Murray, 1880), 281–418.

11. H. G. Wells, David Y. Hughes, and Harry M. Geduld, *A Critical Edition of the War of the Worlds: H. G. Wells's Scientific Romance* (Bloomington: Indiana University Press, 1993), 52. The fate of Tasmania's aborigines was also described in Wells' *Outline of History* (1920).

12. Brantlinger, *Rule of Darkness*, 238, 51.

13. Grant Allen, "Queer Flowers," *Popular Science Monthly* 26, no. 10 (1884): 182.

14. My thanks to Jonathan Smith, whose work brought this aspect of Allen's to my attention. Jonathan Smith, "Grant Allen, Physiological Aesthetics, and the Dissemina-

tion of Darwin's Botany," in *Science Serialized: Representations of the Sciences in Nineteenth-Century Periodicals,* ed. Sally Shuttleworth and Geoffrey N. Cantor (Cambridge, MA: MIT Press, 2004); Jonathan Smith, "Une Fleur du Mal? Swinburne's 'the Sundew' and Darwin's Insectivorous Plants," *Victorian Poetry* 41 (2003): 131–50; and Jonathan Smith, *Charles Darwin and Victorian Visual Culture,* Cambridge Studies in Nineteenth-Century Literature and Culture (Cambridge: Cambridge University Press, 2009). See: Grant Allen, *Charles Darwin* (New York: D. Appleton, 1885).

15. Herbert George Wells, *Experiment in Autobiography. Discoveries and Conclusions of a Very Ordinary Brain (since 1866).* (Philadelphia: J. B. Lippincott, 1934 [1967]), 461. See also: David Y. Hughes, "A Queer Notion of Grant Allen's," *Science Fiction Studies* 25, no. 2 (1998): 271–84; David Cowie, "The Evolutionist at Large: Grant Allen, Scientific Naturalism and Victorian Culture" (PhD diss., University of Kent, 2000); Patrick Parrinder, "The Old Man and His Ghost: Grant Allen, H. G. Wells and Popular Anthropology," in *Grant Allen: Literature and Cultural Politics at the Fin de Siècle,* ed. William Greenslade and Terence Rodgers (Aldershot: Ashgate, 2005).

16. Strangely, there's still no single good overview of Darwin's botany, but if you want to know more try: Allan, *Darwin and His Flowers;* Peter Ayres, *The Aliveness of Plants: The Darwins at the Dawn of Plant Science* (London: Pickering and Chatto, 2008); Richard Bellon, "Charles Darwin Solves the 'Riddle of the Flower'; or, Why Don't Historians of Biology Know about the Birds and the Bees?" *History of Science* 47 (2009): 373–406; Jim Endersby, *A Guinea Pig's History of Biology: The Plants and Animals Who Taught Us the Facts of Life* (London: William Heinemann, 2007); and David Kohn, "Darwin's Botanical Research," in *Charles Darwin at Down House,* ed. Solene Morris, Louise Wilson, and David Kohn (Swindon: English Heritage, 2003).

17. Quoted in: Jonathan Smith, "Grant Allen," 294.

18. Allen, "Queer Flowers," 183.

19. Allen, *Colin Clout's Calendar: The Record of a Summer, April–October* (London: Chatto & Windus, 1883), 98.

20. Ibid., 137.

21. When the orchid *Aracamunia liesneri* was named by named by Germán Carnevali and Ivón Ramirez, they noted its curious glands *and* commented that further work was need to decide whether they "perhaps have an assimilative function." See: J. Steyermark and B. Holst, "Flora of the Venezuelan Guayana, VII: Contributions to the Flora of the Cerro Aracamuni, Venezuela," *Annals of the Missouri Botanical Garden* 76 (1989): 962–64.

22. For example: Arthur Conan Doyle, "The American's Tale" (1880), in Arment, *Flora Curiosa,* 38–45; Manly Wade Wellman, "Come into My Parlour" (1949), in Cassaba, *Roots of Evil,* 73–85; Howard R. Garis, "Professor Jonkin's Cannibal Plant" (1905), in Arment, *Flora Curiosa,* 113–22.

23. *Botanica Delira,* 188–94; *Flora Curiosa,* 188–204; *Botanica Delira,* 273–77; Wyatt Blassingame, "Passion Flower," in *The Unholy Goddess and Other Stories: The Weird Tales of Wyatt Blassingame,* Vol. 3, 13–46 (Vancleave, MS: Dancing Tuatara, 1936 [2011]); Marvin Dana, *The Woman of Orchids* (London: Anthony Treherne, 1901). Oscar Cook's "Si Urag of

the Tail" is included in early editions of Arment, *Flora Curiosa*, but not later ones; it originally appeared in *Weird Tales* 8, no. 1 (1926).

24. Cassaba, *Roots of Evil*, 13–24.

25. Allen, *Colin Clout's Calendar*, 97.

26. Darwin, *Orchids*, 230.

27. Ibid., 87–88.

28. Tina Gianquitto, "Criminal Botany: Progress, Degeneration and Darwin's *Insectivorous Plants*," in *America's Darwin: Darwinian Theory and U. S. Literary Culture*, ed. Tina Gianquitto and Lydia Fisher (Athens: University of Georgia Press, 2014), 235–64.

29. Alfred W. Bennett, "Insectivorous Plants," *Nature* 12, no. 299 (1875): 207.

30. Charles Darwin, *On the Movements and Habits of Climbing Plants* (London: Longman, Green, Longman, Roberts & Green, 1865), 115–18.

31. C. Darwin to W. E. Darwin, [July 25, 1863]: Darwin Correspondence Database, http://www.darwinproject.ac.uk/entry-4199. See also: Endersby, *A Guinea Pig's History of Biology*, 29–60.

32. Charles Darwin, *Insectivorous Plants* (London: John Murray, 1875), 286–320.

33. John Ellor Taylor, ed., *The Sagacity and Morality of Plants: A Sketch of the Life and Conduct of the Vegetable Kingdom . . . New edition, etc.* (London: George Routledge and Sons, 1884 [1904]), v.

34. Ibid., 1–2. He was quoting: Charles Darwin, *The Power of Movement in Plants*, 573.

35. Taylor, *Sagacity and Morality of Plants*, 2–9, 15–16, 88–89.

36. Ibid., 183, 206.

37. Ibid., 61.

38. Mordecai Cubitt Cooke, *Freaks and Marvels of Plant Life: Or, Curiosities of Vegetation* (London: Society for Promoting Christian Knowledge, 1881), 49.

39. Modern historians and philosophers of science are much more dubious, but that is perhaps a subject for a different book.

40. Cooke, *Marvels of Plant Life*, 20.

41. Tegetmeier, "Darwin on Orchids."

42. Allen, "Queer Flowers," 182.

43. Jim Endersby, Introduction to Charles Darwin, *On the Origin of Species by Means of Natural Selection: Or the Preservation of Favoured Races in the Struggle for Life*, ed. Jim Endersby (Cambridge: Cambridge University Press, 2009), l–lvi.

44. Jonathan Smith, "Grant Allen," 288.

45. Allen, "Evolutionist at Large," pp. 36–37. Quoted in: ibid., 292.

46. T. F. H., *Feeble Faith, a Story of Orchids* (London: Hodder and Stoughton, 1882), 4–5, 13.

Chapter 8

1. Ruskin, *Prosperina* (1875–86), quoted in: Martha Hoffman Lewis, "Power & Passion: The Orchid in Literature," in *Orchid Biology: Reviews and Perspectives*, ed. Joseph Arditti (Portland, OR: Timber, 1990), 212–13.

2. According to the *Oxford English Dictionary*, the word first appeared in J. Langhorne's "Sun-flower & Ivy" (in *Fables of Flora* [1771]), "Go, splendid sycophant! no more Display thy soft seductive arts!"

3. The *Oxford English Dictionary* gives 1906 as the first use of "man-eater" and 1911 for "vamp."

4. Arment, *Flora Curiosa*, 188–204.

5. *Botanica Delira*, 283–89; Mike Ashley and Robert A. W. Lowndes, *The Gernsback Days: A Study of the Evolution of Modern Science Fiction* (Rockville, MD: Wildside, 2004).

6. "The Largest Flower in the World," *Amazing Stories*, September 1927, p. 529.

7. Dana, *Woman of Orchids*; "Dana, Marvin Hill," in *Men of Vermont: An Illustrated Biographical History of Vermonters and Sons of Vermont*, ed. Jacob G. Ullery et al. (Brattleboro: Transcript Publishing Co., 1894), 93–94.

8. Ignatz F. Förstermann collected for Sander and Sons from c.1880–86 and was later manager of their branch at Summit, New Jersey. See: Steenis and Steenis-Kruseman, *Flora Malesiana*, Series 1, Vol. 1, 168. See also: http://plants.jstor.org/stable/history/10.5555/al.ap .person.bm000357247

9. Charlotte, "Most Rare."

10. [Rolfe, Robert Allen?], "Dies Orchidianae [Demon Flowers]," *Orchid Review* 4, no. 44 (1896): 233.

11. Edna Worthley Underwood, "An Orchid of Asia: A Tale of the South Seas" (1920); Gordon Philip England, "White Orchids" (1927); both in: Arment, *Arboris Mysterius*, 82–114, 219–29.

12. Sadly, once the orchids disappear from the novel, it degenerates into a rather tiresome and melodramatic morality tale, with Madame Marcou successfully begging forgiveness from her estranged husband, who has killed Arsdale in a duel.

13. Blassingame, *Unholy Goddess*.

14. Underwood, "An Orchid of Asia," in: Arment, *Arboris Mysterius*, 93–99.

15. Dawn Sanders, "Carnivorous Plants: Science and the Literary Imagination," *Planta Carnivora* 32, no. 1 (2010): 30–34. The word seems to derive from "tippet" (a fur collar, or muff), and "twitchy."

16. Smith, "Une Fleur du Mal?"

17. Underwood, "An Orchid of Asia," in: Arment, *Arboris Mysterius*, 91–92.

18. Jane Rendall, "The Citizenship of Women and the Reform Act of 1867," in *Defining the Victorian Nation: Class, Race, Gender and the Reform Act of 1867*, ed. Catherine Hall, Keith McClelland, and Jane Rendall (Cambridge: Cambridge University Press, 2000), 136–38.

19. St. Louis (MO) *Globe-Democrat*, July 21, 1879, *Oxford English Dictionary*.

20. Rebecca Stott, *Fabrication of the Late Victorian Femme Fatale: The Kiss of Death* (Basingstoke: Macmillan, 1992), viii–xiii, 1–30.

21. Sally Ledger and Roger Luckhurst, eds., *The Fin de Siècle: A Reader in Cultural History, c.1880–1900* (Oxford: Oxford University Press, 2000), 75–96.

22. Lucy Bland, "The Married Woman, the 'New Woman' and the Feminist: Sexual

Politics in the 1890s," in *Equal or Different: Women's Politics, 1800–1914*, ed. Jane Rendall (Oxford: Blackwell, 1987), 141–62.

23. William Greenslade and Terence Rodgers, eds., *Grant Allen: Literature and Cultural Politics at the Fin de Siècle* (Aldershot: Ashgate, 2005), 13.

24. Lewis, "Orchid in Literature," 217.

25. Ibid., 214–15.

26. Oscar Wilde, *The Picture of Dorian Gray* (Harmondsworth: Penguin, 1891 [1985]).

27. Joris-Karl Huysmans, *Against Nature* (*À Rebours*) (Harmondsworth: Penguin, 1884 [1959]).

28. Marcel Proust, *Swann's Way*, trans. C. K. Scott Moncrieff and Terence Kilmartin, Vol. 1 of *In Search of Lost Time* (London: Vintage, 1913 [2005]).

29. James Neil, *Rays from the Realms of Nature: Or, Parables of Plant Life*, 4th ed. (London: Cassell, Petter, Galpin, 1879 [1884]).

30. Ibid., 25–27.

31. I wonder if Ian Fleming knew this text and whether the author's middle name inspired him to name *Moonraker*'s orchid-wielding villain, Hugo Drax.

32. Lewis, "Orchid in Literature," 218.

33. Richard Le Gallienne, "Fractional Humanity" (1894), in: ibid.

34. Ashmore Russan and Frederick Boyle, *The Orchid Seekers: A Story of Adventure in Borneo* (London: Frederick Warne, 1897).

35. Even Sanders' biography, which veers towards outright hagiography, notes that Boyle was regularly criticized for his inaccurate dates and exaggerated stories. Swinson, *Frederick Sander*, 16–20.

36. Charlotte, "Most Rare."

37. Boyle, *About Orchids*, 145–46.

38. Ashmore Russan and Frederick Boyle, *The Riders; or, through Forest and Savannah with the Red Cockades*, ed. Frederick Boyle (London: Frederick Warne, 1896).

39. Percy Ainslie, *The Priceless Orchid: A Story of Adventure in the Forests of Yucatan* (London: Sampson Low, Marston, 1892). Anon. review of *The Priceless Orchid* in *Spectator*, December 24, 1892, 933.

40. Henry Rider Haggard, *Allan and the Holy Flower* (Berkeley Heights, NJ: Wildside, 1915 [1999]). For Haggard's own gardening, see: *A Gardener's Year* (London: Longmans, Green, 1905).

41. Haggard, *Holy Flower*, 46.

42. Russan and Boyle, *Orchid Seekers*, 15.

Chapter 9

1. *The Big Sleep* (1939), in: Raymond Chandler, *Three Novels* (Harmondsworth: Penguin Books, 1993), 6.

2. Charles J. Rzepka, "'I'm in the Business Too': Gothic Chivalry, Private Eyes, and Proxy Sex and Violence in Chandler's *The Big Sleep*," *Modern Fiction Studies* 46, no. 3 (2000): 708–9. See also: Ernest Fontana, "Chivalry and Modernity in Raymond Chandler's

The Big Sleep," in *The Critical Response to Raymond Chandler,* ed. J. K. Van Dover (Westport: Greenwood, 1995), 159–65.

3. The distinguished orchid biologist Joseph Arditti, for example, wondered why Chandler introduced them only so they could be "vilified." Joseph Arditti, "Orchids in Novels, Music, Parables, Quotes, Secrets, and Odds and Ends," *Orchid Review* 92 (1984): 373.

4. Rex Stout, *The League of Frightened Men* (New York: Bantam Books, 1935 [1985]), 4–5.

5. John H. Vandermeulen, "Nero Wolfe — Orchidist Extraordinaire," *American Orchid Society Bulletin* 54, no. 2 (1985): 146.

6. Letter to Mrs. Dorothy Beach, June 3, 1957: Frank MacShane, ed., *Selected Letters of Raymond Chandler* (London: Cape, 1982), 453.

7. Vandermeulen, "Nero Wolfe."

8. http://www.tcm.com/mediaroom/video/69968/Brother-Orchid-Original-Trailer-.html

9. http://www.imdb.com/title/tt0026829/?ref_=fn_al_tt_1; Lewis, "Orchid in Literature," 228–29; *The Big Sleep,* in: Chandler, *Three Novels,* 136.

10. Jocelyn Brooke, *The Wild Orchids of Britain* (London: Bodley Head, 1950), 15.

11. *The Orchid Trilogy* (Harmondsworth: Penguin Books, 1981). All ellipses in the quotes that follow are Brooke's. Jonathan Hunt's introduction to the 2002 Penguin Classics edition of *The Military Orchid* was especially useful in understanding Brooke's work; see: http://jocelynbrooke.com/an-introduction-to-the-military-orchid/

12. Strictly speaking, this plot twist is wrong, because fir bark and other cheaper materials had already begun to replace *Osmunda* in the 1950s, but why let historical pedantry spoil a good story? See: Reinikka, *History of the Orchid,* 57.

13. Interestingly, there's a widespread myth that Poitier improvised the slap, hence Larry Gates' surprise, but according to Jewison, not only was the retaliatory slap scripted, but Poitier insisted it be in every print of the film, in every market where it was distributed. See: http://www.dga.org/Craft/DGAQ/All-Articles/1101-Spring-2011/Shot-to-Remember-Norman-Jewison.aspx

Chapter 10

1. Darwin, *Orchids,* 292.

2. Gerard Edwards Smith, *Catalogue of Rare or Remarkable Phaenogamous Plants, Collected in South Kent* (London: Longman, Rees, Orme Brown, and Green, 1829), 52.

3. Darwin, *Orchids,* 68.

4. M. Åsberg and W. T. Stearn, "Linnaeus's Öland and Gotland Journey 1741," *Biological Journal of the Linnean Society* 5, no. 1 (1973): 1–220; Jarvis and Cribb, "Linnaean Concepts of Orchids," 369.

5. Darwin, *Orchids,* 46–47.

6. E. G. Britton, "The Swiss League for the Protection of Nature," *Torreya* 19, no. 5 (1919).

7. Oakes Ames, *Pollination of Orchids through Pseudocopulation (Botanical Museum Leaflets, Vol. V, No. 1)* (Cambridge, MA: Botanical Museum of Harvard University, 1937), 3.

8. Correvon and Pouyanne, "Un Curieux Cas de Mimétisme chez les Ophrydées," *Journal de la Société Nationale d'Horticulture de France*, Series 4, no. 17 (1916): 42. Translated from the French by Nicholas J. Vereecken and Ana Francisco, "Ophrys Pollination: From Darwin to the Present Day," in *Darwin's Orchids: Then and Now*, ed. Retha Edens-Meier and Peter Bernhardt (Chicago: University of Chicago Press, 2014), 53.

9. Ames, *Pseudocopulation*, 6–7.

10. Correvon and Pouyanne, "Un Curieux Cas de Mimétisme chez les Ophrydées," *Journal de la Société Nationale d'Horticulture de France*, Series 4, no. 17 (1916): 42. Quoted in: ibid., 6.

11. Edith Coleman, "Pollination of the Orchid *Cryptostylis leptochila*," *Victorian Naturalist* 44, no. 1 (1927).

12. "[Obituary] Col. M. J. Godfery," *Nature* 155, no. 3943 (1945).

13. "Pollination of the Orchid *Cryptostylis leptochila*," *Victorian Naturalist* 44, no. 532 (1928): 334–37, 40.

14. Ames, *Pseudocopulation*, 12; Allan McEvey, "Coleman, Edith," *Australian Dictionary of Biography*, http://adb.anu.edu.au/biography/coleman-edith-9784/text17291

15. Vereecken and Francisco, "Ophrys Pollination," 56–57.

16. Coleman, "Pollination of the Orchid *Cryptostylis leptochila*," 336–37, 40.

17. Stearn, "Two Thousand Years of Orchidology," 27–28.

18. Ames, *Pseudocopulation*.

19. Ibid., 18; the final phrase is from an early-nineteenth century hymn, "From Greenland's Icy Mountains," also known as the "Missionary Hymn," by Reginald Heber.

20. https://www.darwinproject.ac.uk/darwins-notes-on-marriage. See: Evelleen Richards, "Darwin and the Descent of Women," in *The Wider Domain of Evolutionary Thought*, ed. David Oldroyd and Ian Langham (Dordrecht: D. Reidel, 1983), 87; Endersby, "Darwin on Generation, Pangenesis and Sexual Selection."

21. I explore this connection and the wider history of pseudocopulation's discovery in more detail in "Deceived by Orchids: Sex, Science, Fiction and Darwin," *British Journal for the History of Science* 49 (June 2016): 205–29.

22. Coleman's grandson, Peter, remembers her reading a lot of Wells' books. Personal communication, via Danielle Clode, Flinders University, September 1, 2015.

23. David Seed, ed., *A Companion to Science Fiction* (Oxford: Blackwell, 2005), 39.

24. John Collier, "Green Thoughts," in *Roots of Evil: Beyond the Secret Life of Plants*, ed. Carlos Cassaba (London: Corgi, 1931 [1976]), 121–36.

25. Arthur C. Clarke, "The Reluctant Orchid," in *Tales from the White Hart* (New York: Ballantine, 1956 [1957]), 103–13.

26. J. G. Ballard, "Prima Belladona," in *The Four-Dimensional Nightmare* (London: Science Fiction Book Club, 1956 [1963]), 79–92. Ballard published another SF story, "Escapement," in the same month in *New Worlds* (December 1956); see the Internet Speculative Fiction Database (http://www.isfdb.org/) for details of both.

27. Herbert W. Franke, *The Orchid Cage*, trans. Christine Priest (New York: Daw, 1961 [1973]).

28. Pete Adams and Charles Nightingale, "Planting Time," in *Galactic Empires*, Vol. 1, ed. Brian W. Aldiss (New York: St Martin's, 1976).

29. John Boyd, *The Pollinators of Eden* (London: Pan Science Fiction, 1969 [1972]). Boyd was the main pen-name of Boyd Bradfield Upchurch.

30. See: Brantlinger, *Rule of Darkness*, 234–47.

31. John Wyndham, *Day of the Triffids* (Harmondsworth: Penguin, 1951 [1954]), 32–33; Bernhardt, *Wily Violets*, 210–11.

Chapter 11

1. Leonard Huxley, *Life and Letters of Joseph Dalton Hooker*, Vol. 1 (London: John Murray, 1918), 337.

2. Joseph Dalton Hooker, *Himalayan Journals: Or, Notes of a Naturalist in Bengal, the Sikkim and Nepal Himalayas, the Khasia Mountains*, Vol. 2 (London: John Murray, 1855), 321–22.

3. Williams, *Orchid-Grower's Manual*, 2–6.

4. Albert Millican, *Travels and Adventures of an Orchid Hunter: An Account of Canoe and Camp Life in Colombia, While Collecting Orchids in the Northern Andes* (London: Cassell, 1891), vii.

5. Boyle, *Woodlands Orchids* 144–45.

6. "Observations on the Capacity of Orchids to Survive in the Struggle for Existence," *Orchid Review* 30 (1922): 229–234. Collected in: Ames, *Orchids in Retrospect*, 12–13.

7. Joseph Arditti, "An History of Orchid Hybridization, Seed Germination and Tissue Culture," *Botanical Journal of the Linnean Society* 89 (1984): 359–81. Many other plants also rely on mycorrhiza, which were first named in 1885 by Albert Bernhard Frank. Albert Bernhard Frank and James M. Trappe, "On the Nutritional Dependence of Certain Trees on Root Symbiosis with Belowground Fungi [an English translation of A. B. Frank's classic paper of 1885]," *Mycorrhiza* 15 (2004): 267–75.

8. As my undergraduate biology lecturer, Professor Michael Archer, said after mentioning this fact: if you remember nothing else from today, remember this — never take a nap under a Seychelles Double Coconut tree.

9. Arditti, "History of Orchid Hybridization," 370–72.

10. Stearn, "Two Thousand Years of Orchidology," 28.

11. Bernhardt, *Wily Violets*, 192–93.

12. H. Reinheimer, *Symbiosis: A Socio-Physiological Study of Evolution* (London: Headley Brothers, 1920), 243–44. For more on Reinheimer, see: Jan Sapp, *Evolution by Association: A History of Symbiosis* (New York: Oxford University Press, 1994), 60–61.

13. Ames, *Orchids in Retrospect*, 14–15; Douglas Houghton Campbell, "The New Flora of Krakatau," *American Naturalist* 43, no. 512 (1909): 449–60.

14. Information in this section comes from interviews with Michael Hutchings and I am most grateful for his assistance.

Conclusion

1. http://www.iucnredlist.org/

2. Arditti, "History of Orchid Hybridization," 371–72.

3. James Agee, "The U.S. Commercial Orchid," in *James Agee: Selected Journalism*, ed. Paul Ashdown (Knoxville: University of Tennessee Press, 1985 [2005]). A curious coincidence: Agee rewrote the screenplay for the movie *The African Queen*, the first version of which was by John Collier, who wrote "Green Thoughts" (see chapter 10).

4. Don Wharton, "An Orchid a Minute," *Saturday Evening Post* 213, no. 41 (1941): 16–84.

5. Agee to Father Flye, August 23, 1935: James Agee, *Letters of James Agee to Father Flye* (New York: George Braziller, 1962), 77.

6. Agee to Father Flye, September 17, 1935: ibid., 81–82.

7. Ibid.

8. S. M. Walters, "The Name of the Rose: A Review of Ideas on the European Bias in Angiosperm Classification," *New Phytologist*, no. 104 (1986): 527–46.

9. Lynda H. Schneekloth, "Plants: The Ultimate Alien," *Extrapolations* 42, no. 3 (2001): 246–54.

10. Agee, "Commerical Orchid," 110. Agee didn't name him, but the *Saturday Evening Post* gave his name as Godfrey Erickson. Wharton, "Orchid a Minute," 82.

Bibliography

Adams, Pete, and Charles Nightingale. "Planting Time." In *Galactic Empires*, Vol. 1, edited by Brian W. Aldiss, 293–305. New York: St. Martin's, 1976.

Agee, James. *Letters of James Agee to Father Flye*. New York: George Braziller, 1962.

———. "The U.S. Commercial Orchid." In *James Agee: Selected Journalism*, edited by Paul Ashdown. Knoxville: University of Tennessee Press, 1985 (2005).

Ainslie, Percy. *The Priceless Orchid: A Story of Adventure in the Forests of Yucatan.* London: Sampson Low, Marston, 1892.

Allan, Mea. *Darwin and His Flowers: The Key to Natural Selection*. London: Faber and Faber, 1977.

Allen, Grant. *Charles Darwin*. New York: D. Appleton, 1885.

———. *Colin Clout's Calendar: The Record of a Summer, April–October*. London: Chatto & Windus, 1883.

———. "Queer Flowers." *Popular Science Monthly* 26, no. 10 (1884): 177–87.

Ames, Oakes. *Orchids in Retrospect: A Collection of Essays on the Orchidaceae*. Cambridge, MA: Botanical Museum of Harvard University, 1948.

———. *Pollination of Orchids through Pseudocopulation (Botanical Museum Leaflets, Vol. V, No. 1)*. Cambridge, MA: Botanical Museum of Harvard University, 1937.

Anonymous. "Fertilization of Orchids. By Charles Darwin." *Parthenon* 1, no. 6 (1862): 177–78.

———. "Mr. Darwin's Orchids." *Saturday Review of Politics, Literature, Science, and Art* (1862): 486.

Arber, Agnes. *Herbals: Their Origin and Evolution. A Chapter in the History of*

Botany 1470–1670. 3rd ed. Facsimile of the 1938 2nd edition, with an introduction by W. T. Stearn. Cambridge: Cambridge University Press, 1990.

Arditti, Joseph. "An History of Orchid Hybridization, Seed Germination and Tissue Culture." *Botanical Journal of the Linnean Society* 89 (1984): 359–81.

———. "Orchids in Novels, Music, Parables, Quotes, Secrets, and Odds and Ends." *Orchid Review* 92 (1984): 373–76.

Arditti, Joseph, John Elliott, Ian J. Kitching, and Lutz T. Wasserthal. "'Good Heavens What Insect Can Suck It'—Charles Darwin, *Angraecum sesquipedale* and *Xanthopan morganii praedicta.*" *Botanical Journal of the Linnean Society* 169 (2012): 403–32.

Arment, Chad, ed. *Arboris Mysterius: Stories of the Uncanny and Undescribed from the Botanical Kingdom.* Greenville, OH: Coachwhip, 2014.

———, ed. *Botanica Delira: More Stories of Strange, Undiscovered, and Murderous Vegetation.* Landisville, PA: Coachwhip, 2010.

———, ed. *Flora Curiosa: Cryptobotany, Mysterious Fungi, Sentient Trees, and Deadly Plants in Classic Science Fiction and Fantasy.* 2nd (rev.) ed. Greenville, OH: Coachwhip, 2013.

Armstrong, Isobel. *Victorian Glassworlds: Glass Culture and the Imagination 1830–1880.* Oxford: Oxford University Press, 2008.

Ashley, Mike, and Robert A. W. Lowndes. *The Gernsback Days: A Study of the Evolution of Modern Science Fiction.* Rockville, MD: Wildside, 2004.

Auerbach, Jeffrey A. *The Great Exhibition of 1851: A Nation on Display.* New Haven: Yale University Press, 1999.

Ayres, Peter. *The Aliveness of Plants: The Darwins at the Dawn of Plant Science.* London: Pickering and Chatto, 2008.

Ballard, J. G. "Prima Belladona." In *The Four-Dimensional Nightmare,* 79–92. London: Science Fiction Book Club, 1956 (1963).

Bateman, James. *The Orchidaceae of Mexico and Guatemala.* London: For the author, J. Ridgway & Sons, 1837–1843.

Bauer, Ralph. "A New World of Secrets: Occult Philosophy and Local Knowledge in the Sixteenth-Century Atlantic." In *Science and Empire in the Atlantic World,* edited by James Delbourgo and Nicholas Dew, 99–126. London: Routledge, 2008.

Beatty, John. "Chance Variation: Darwin on Orchids." *Philosophy of Science* 73, no. 5 (2006): 629–41.

Bellon, Richard. "Charles Darwin Solves the "Riddle of the Flower"; or, Why Don't Historians of Biology Know about the Birds and the Bees?" *History of Science* 47 (2009): 373–406.

———. "Inspiration in the Harness of Daily Labor: Darwin, Botany, and the Triumph of Evolution, 1859–1868." *Isis* 102 (2011): 393–420.

Bennett, Alfred W. "Insectivorous Plants." *Nature* 12, no. 299 (1875): 228–31.

Bernhardt, Peter. *Gods and Goddesses in the Garden: Greco-Roman Mythology and the Scientific Names of Plants.* New Brunswick, NJ: Rutgers University Press, 2008.

———. *Wily Violets and Underground Orchids: Revelations of a Botanist.* New York: Vintage, 1990.

Bland, Lucy. "The Married Woman, the 'New Woman' and the Feminist: Sexual Politics in the 1890s." In *Equal or Different: Women's Politics, 1800–1914*, edited by Jane Rendall. Oxford: Blackwell, 1987.

Blassingame, Wyatt. "Passion Flower." In *The Unholy Goddess and Other Stories: The Weird Tales of Wyatt Blassingame*, Vol. 3, 13–46. Vancleave, MS: Dancing Tuatara, 1936 (2011).

Blumenfeld-Kosinski, Renate. "The Scandal of Pasiphae: Narration and Interpretation in the 'Ovide Moralisé.'" *Modern Philology* 93, no. 3 (1996): 307–26.

Blunt, Wilfrid. *The Compleat Naturalist: A Life of Linnaeus*. London: Collins, 1971.

Boyd, John. *The Pollinators of Eden*. London: Pan Science Fiction, 1969 (1972).

Boyle, Frederick. *About Orchids: A Chat*. London: Chapman & Hall, 1893.

———. "The Story of an Orchid." *Ludgate* 4 (1897): 508–11.

———. *The Woodlands Orchids Described and Illustrated, with Stories of Orchid-Collecting*. London: Macmillan, 1901.

Boyle, Frederick, and Joseph Godseff. *The Culture of Greenhouse Orchids: Old System and New*. London: Chapman & Hall, 1902.

Brantlinger, Patrick. *Rule of Darkness: British Literature and Imperialism, 1830–1914*. Ithaca: Cornell University Press, 1988.

Britton, E. G. "The Swiss League for the Protection of Nature." *Torreya* 19, no. 5 (1919): 101–2.

Brooke, James. *The Fairfield Orchids, a Descriptive Catalogue of the Species and Varieties Grown by James Brooke and Co., Fairfield Nurseries*. Manchester: James Brooke, 1872.

Brooke, Jocelyn. *The Orchid Trilogy*. Harmondsworth: Penguin, 1981.

———. *The Wild Orchids of Britain*. London: Bodley Head, 1950.

Brooke, John Hedley. *Science and Religion: Some Historical Perspectives*. Cambridge: Cambridge University Press, 1991.

Browne, Janet. "Botany for Gentlemen: Erasmus Darwin and *The Loves of the Plants*." *Isis* 80, no. 304 (1989): 593–621.

Burkhardt, Frederick, Duncan M. Porter, Janet Browne, and Marsha Richmond, eds. *The Correspondence of Charles Darwin. Vol. 8, 1860*. Cambridge: Cambridge University Press, 1993.

Burkhardt, Frederick, James A. Secord, and the Editors of the Darwin Correspondence Project, eds. *The Correspondence of Charles Darwin. Vol. 24, 1874*. Cambridge: Cambridge University Press, 2015.

Buzard, James, Joseph W. Childers, and Eileen Gillooly, eds. *Victorian Prism: Refractions of the Crystal Palace*. Charlottesville: University of Virginia Press, 2007.

Cain, Arthur James. "Rank and Sequence in Caspar Bauhin's Pinax." *Botanical Journal of the Linnean Society* 114 (1994): 311–56.

Campbell, Douglas Houghton. "The New Flora of Krakatau." *American Naturalist* 43, no. 512 (1909): 449–60.

Campbell, George (8th Duke of Argyll). "The Supernatural [Review of Darwin on Orchids and Other Works]." *Edinburgh Review* 116, no. 236 (1862): 236–45.

———. "What Is Science?" *Good Words* 26 (1885): 236–45.

Cañizares-Esguerra, Jorge. "Iberian Science in the Renaissance: Ignored How Much Longer?" *Perspectives on Science* 12, no. 1 (2004): 86–124.

Carnevali, Germán, and Ivón Ramírez. "New or Noteworthy Orchids for the Venezuelan Flora, VII: Additions in Maxillaria from the Venezuelan Guayana." *Annals of the Missouri Botanical Garden* 76, no. 2 (1989): 374–80.

Cassaba, Carlos, ed. *Roots of Evil: Beyond the Secret Life of Plants*. London: Corgi, 1976.

Chandler, Raymond. *Three Novels*. Harmondsworth: Penguin, 1993.

Charlotte, Lady. "Most Rare: Flowers That Cost Lives to Secure." *Daily Mail*, no. 3 (1896): 7.

Clarke, Arthur C. "The Reluctant Orchid." In *Tales from the White Hart*, 103–13. New York: Ballantine, 1956 (1957).

Coats, Alice M. *The Quest for Plants: A History of Horticultural Explorers*. London: Studio Vista, 1969.

Coe, Sophie D., and Michael D. Coe. *The True History of Chocolate*. London: Thames & Hudson, 1996 (2013).

Coleman, Edith. "Pollination of the Orchid *Cryptostylis leptochila*." *Victorian Naturalist* 44, no. 1 (1927): 20–22.

———. "Pollination of the Orchid *Cryptostylis leptochila*." *Victorian Naturalist* 44, no. 532 (1928): 333–40.

Collier, John. "Green Thoughts." In *Roots of Evil: Beyond the Secret Life of Plants*, edited by Carlos Cassaba, 121–36. London: Corgi, 1931 (1976).

Collins, Percy. "The Romance of Auctioneering." *Strand Magazine* 31, no. 181 (1906): 100–106.

Cooke, Mordecai Cubitt. *Freaks and Marvels of Plant Life: Or, Curiosities of Vegetation*. London: Society for Promoting Christian Knowledge, 1881.

Correll, Donovan S. "Vanilla: Its Botany, History, Cultivation and Economic Import." *Economic Botany* 7, no. 4 (1953): 291–358.

Cowie, David. "The Evolutionist at Large: Grant Allen, Scientific Naturalism and Victorian Culture." PhD diss., University of Kent, 2000.

Cribb, Phillip, Mike Tibbs, John Day, and Kew Royal Botanic Gardens. *A Very Victorian Passion: The Orchid Paintings of John Day, 1863 to 1888*. Kew: Royal Botanic Garden, 2004.

Croll, Oswald, and Johann Hartmann. *A Treatise of Oswaldus Crollius of Signatures of Internal Things; or, A True and Lively Anatomy of the Greater and Lesser World*. London: Starkey, 1669.

Cruz, Martin de la. *The Badianus Manuscript: (Codex Barberini, Latin 241) Vatican Library; An Aztec Herbal of 1552*. Baltimore: Johns Hopkins University Press, 1940.

Dalby, Andrew. "The Name of the Rose Again; or, What Happened to Theophrastus 'On Aphrodisiacs'?" *Petits Propos Culinaires* 64 (2000): 9–15.

Dana, Marvin. *The Woman of Orchids*. London: Anthony Treherne, 1901.

Darwin, Charles. *Autobiographies*. Edited by Michael Neve and Sharon Messenger. Harmondsworth: Penguin, 1887/1903 (2003).

————. *Insectivorous Plants*. London: John Murray, 1875.

————. *On the Movements and Habits of Climbing Plants*. London: Longman, Green, Longman, Roberts & Green, 1865.

————. *On the Origin of Species by Means of Natural Selection: Or the Preservation of Favoured Races in the Struggle for Life*. London: John Murray, 1859.

————. *On the Various Contrivances by Which British and Foreign Orchids Are Fertilised by Insects, and on the Good Effects of Intercrossing*. London: John Murray, 1862.

————. *The Power of Movement in Plants*. London: John Murray, 1880.

————. *The Various Contrivances by Which British and Foreign Orchids Are Fertilised by Insects, and on the Good Effects of Intercrossing*. 2nd ed. London: John Murray, 1877 (1904).

Darwin, Charles, and Francis Darwin. *The Power of Movement in Plants*. London: John Murray, 1880.

Darwin, Francis. *The Life and Letters of Charles Darwin*. Rev. ed. Vol. 3. London: John Murray, 1888.

de Laet, Johannes. "Manuscript Translation of Hernández." Translated by Rafael Chabrán, Cynthia L. Chamberlin, and Simon Varey. In *The Mexican Treasury: The Writings of Dr. Francisco Hernández*, edited by Simon Varey, 161–72. Stanford: Stanford University Press, 2000.

Dear, Peter. *Revolutionizing the Sciences: European Knowledge and Its Ambitions, 1500–1700*. Basingstoke: Palgrave, 2001.

Debus, Allen G. *The Chemical Philosophy: Paracelsian Science and Medicine in the Sixteenth and Seventeenth Centuries*. New York: Science History, 1977.

Dioscorides, Pedianus. *De Materia Medica*. Edited by Tess Anne Osbaldeston. Johannesburg, South Africa: Ibidis, 2000. Original is from ca. 50–60 CE.

Dioscorides, Pedanius. *De Materia Medica*. Translated by Lily Y. Beck. New York: Olms-Weidmann, 2005. Original is from ca. 50–60 CE.

Dixon, Thomas. *Science and Religion: A Very Short Introduction*. Oxford: Oxford University Press, 2008.

Drayton, Richard H. *Nature's Government: Science, Imperial Britain and the "Improvement" of the World*. New Haven: Yale University Press, 2000.

Dupree, A. Hunter. *Asa Gray: American Botanist, Friend of Darwin*. Baltimore: Johns Hopkins University Press, 1959 (1988).

Edens-Meier, Retha, and Peter Bernhardt, eds. *Darwin's Orchids: Then and Now*. Chicago: University of Chicago Press, 2014.

Eisenstein, Elizabeth L. *The Printing Revolution in Early Modern Europe*. Cambridge: Cambridge University Press, 1983.

Elliott, Brent. "The Royal Horticultural Society and Its Orchids: A Social History." *Occasional Papers from the RHS Lindley Library* 2 (2010): 3–53.

Endersby, Jim. "Darwin on Generation, Pangenesis and Sexual Selection." In *Cambridge Companion to Darwin*, edited by M. J. S. Hodge and G. Radick, 69–91. Cambridge: Cambridge University Press, 2003.

————. *A Guinea Pig's History of Biology: The Plants and Animals Who Taught Us the Facts of Life*. London: William Heinemann, 2007.

————. *Imperial Nature: Joseph Hooker and the Practices of Victorian Science*. Chicago: University of Chicago Press, 2008.

————. Introduction to *On the Origin of Species by Means of Natural Selection: Or the Preservation of Favoured Races in the Struggle for Life,* by Charles Darwin, xi–lxv, edited by Jim Endersby. Cambridge: Cambridge University Press, 2009.

Findlen, Paula. "Natural History." In *The Cambridge History of Science.* Vol. 3, *Early Modern Science*, edited by Katharine Park and Lorraine Daston, 435–68. Cambridge: Cambridge University Press, 2006.

Finucci, Valeria. *The Manly Masquerade: Masculinity, Paternity and Castration in the Italian Renaissance*. Durham, NC: Duke University Press, 2003.

Forbes, Henry Ogg. "Three Months' Exploration in the Tenimber Islands, or Timor Laut." *Proceedings of the Royal Geographical Society and Monthly Record of Geography* New Monthly Series 6, no. 3 (1884): 113–29.

Frank, Albert Bernhard, and James M. Trappe. "On the Nutritional Dependence of Certain Trees on Root Symbiosis with Belowground Fungi [an English translation of A. B. Frank's classic paper of 1885]." *Mycorrhiza* 15 (2004): 267–75.

Franke, Herbert W. *The Orchid Cage*. Translated by Christine Priest. New York: Daw, 1961 (1973).

Gerard, John. *The Herball: Or, Generall Historie of Plants* [photoreprint of the 1597 ed. published by J. Norton, London]. The English Experience, Its Record in Early Printed Books Published in Facsimile. 2 vols. Norwood, NJ: W. J. Johnson, 1974.

Ghiselin, Michael T. "Darwin: A Reader's Guide." *Occasional Papers of the California Academy of Sciences* 155 (2009): 1–185.

Gianquitto, Tina. "Criminal Botany: Progress, Degeneration and Darwin's *Insectivorous Plants*." In *America's Darwin: Darwinian Theory and U. S. Literary Culture*, edited by Tina Gianquitto and Lydia Fisher, 235–64. Athens: University of Georgia Press, 2014.

Giesecke, Annette. *The Mythology of Plants: Botanical Lore from Ancient Greece and Rome*. Los Angeles: J. Paul Getty Museum, 2014.

Gimmel, Millie. "Reading Medicine in the *Codex de la Cruz Badiano*." *Journal of the History of Ideas* 69, no. 2 (2008): 169–92.

Gray, Asa, ed. *Darwiniana: Essays and Reviews Pertaining to Darwinism*. Edited by A. Hunter Dupree. Cambridge: Harvard University Press, 1963.

————. "[Review of C. R. Darwin] On the Various Contrivances by Which British and Foreign Orchids Are Fertilised by Insects." *American Journal of Science and Arts* New Series 34 (1862): 138–41.

"The Great Orchid Sale." *Standard*, October 17, 1891.

Greene, Edward Lee. *Landmarks of Botanical History. A Study of Certain Epochs in the Development of the Science of Botany. Part I.—Prior to 1562 A.D.* Smithsonian Miscellaneous Collections. Vol. 54. Washington, DC: Smithsonian Institution Press, 1909.

————. *Landmarks of Botanical History: Part 1*. Publications of the Hunt Institute for

Botanical Documentation Carnegie-Mellon University. Edited by Frank N. Egerton. Vol. 1. Stanford California: Stanford University Press, 1909–1915 (1983).

Greenslade, William, and Terence Rodgers, eds. *Grant Allen: Literature and Cultural Politics at the Fin de Siècle.* Aldershot: Ashgate, 2005.

Haggard, Henry Rider. *Allan and the Holy Flower.* Berkeley Heights, NJ: Wildside, 1915 (1999).

———. *A Gardener's Year.* London: Longmans, Green, 1905.

Harrison, Peter. "Linnaeus as a Second Adam? Taxonomy and the Religious Vocation." *Zygon* 44, no. 4 (2009): 879–93.

Hooker, Joseph Dalton. "*Dendrobium phalaenopsis.*" *Curtis's Botanical Magazine, comprising the plants of the Royal Gardens of Kew and of other botanical establishments in Great Britain.* Third Series 41 (1885): t. 6817.

———. *Himalayan Journals: Or, Notes of a Naturalist in Bengal, the Sikkim and Nepal Himalayas, the Khasia Mountains.* Vol. 2. London: John Murray, 1855.

———. "[Review of] Darwin, C. R. On the Various Contrivances by Which British and Foreign Orchids Are Fertilized by Insects." *Natural History Review* 2, no. 8 (1862): 371–76.

Hughes, David Y. "A Queer Notion of Grant Allen's." *Science Fiction Studies* 25, no. 2 (1998): 271–84.

Huxley, Leonard. *Life and Letters of Joseph Dalton Hooker.* Vol. 1. London: John Murray, 1918.

Huysmans, Joris-Karl. *Against Nature (À Rebours).* Harmondsworth: Penguin, 1884 (1959).

Jarvis, Charlie, and Phillip Cribb. "Linnaean Sources and Concepts of Orchids." *Annals of Botany* 104 (2009): 365–76.

Koerner, Lisbet. "Carl Linnaeus in His Time and Place." In *Cultures of Natural History,* edited by Nicholas Jardine, James A. Secord, and Emma Spary, 145–62. Cambridge: Cambridge University Press, 1996.

———. *Linnaeus: Nature and Nation.* Cambridge, MA: Harvard University Press, 1999.

Kohn, David. "Darwin's Botanical Research." In *Charles Darwin at Down House,* edited by Solene Morris, Louise Wilson, and David Kohn, 50–59. Swindon: English Heritage, 2003.

Kumbaric, Alma, Valentina Savo, and Giulia Caneva. "Orchids in the Roman Culture and Iconography: Evidence for the First Representations in Antiquity." *Journal of Cultural Heritage* 14 (2013): 311–16.

Laertius, Diogenes. *The Lives and Opinions of Eminent Philosophers.* Translated by C. D. Yonge. Bohn's Classical Library. London: Henry G. Bohn, 1853.

Larson, James. "Linnaeus and the Natural Method." *Isis* 58, no. 3 (1967): 304–20.

Latour, Bruno. *Science in Action: How to Follow Scientists and Engineers through Society.* Cambridge, MA: Harvard University Press, 1987 (1994).

Ledger, Sally, and Roger Luckhurst, eds. *The Fin de Siècle: A Reader in Cultural History, c. 1880–1900.* Oxford: Oxford University Press, 2000.

[Leifchild, John Roby]. "Review of Darwin, CR 'on the Various Contrivances by Which

British and Foreign Orchids Are Fertilized by Insects.'" *Athenaeum*, no. 1804 (1862): 683–85.

Lennox, James G. "Darwin *Was* a Teleologist." *Biology and Philosophy* 8, no. 4 (1993): 409–21.

Lewis, Martha Hoffman. "Power & Passion: The Orchid in Literature." In *Orchid Biology: Reviews and Perspectives*, edited by Joseph Arditti, 207–49. Portland, OR: Timber, 1990.

Liger, Louis. *The Compleat Florist: Or, the Universal Culture of Flowers, Trees and Shrubs; Proper to Imbellish Gardens*. London: Printed for Benj. Tooke, at the Temple-Gate, Fleetstreet, 1706.

Lightman, Bernard, and Gowan Dawson, eds. *Victorian Scientific Naturalism: Community, Identity, Continuity*. Chicago: University of Chicago Press, 2014.

Linden, Lucien. "History of the Introduction of *Eulophiella elisabethae*." *Lindenia* 5, no. 29 (1893): 46, 50.

Lindley, John. *Collectanea Botanica, or Figures and Botanical Illustrations of Rare and Curious Exotic Plants*. London: Richard and Arthur Taylor, Shoe-Lane, 1821.

Lloyd, G. E. R. *Science, Folklore and Ideology: Studies in the Life Sciences in Ancient Greece*. Cambridge: Cambridge University Press, 1983.

"London Letter." *Garden and Forest* 6, no. 281 (1893): 294.

Lowood, Henry. "The New World and the European Catalog of Nature." In *America in European Consciousness, 1493–1750*, edited by Karen Ordahl Kupperman, 295–323. Chapel Hill: University of North Carolina Press, 1995.

Lyons, John Charles. *Remarks on the Management of Orchidaceous Plants, with a Catalogue of Those in the Collection of J. C. Lyons, Ladiston. Alphabetically Arranged, with Their Native Countries and a Short Account of the Mode of Cultivation Adopted*. Ladiston: Self-published, 1843.

"M. Hamelin's Adventures in Madagascar." *Standard*, July 25, 1893, 6.

MacShane, Frank, ed. *Selected Letters of Raymond Chandler*. London: Cape, 1982.

"Madagascar: — *Eulophiella elisabethae*." *Gardeners' Chronicle* Third Series 14, no. 358 (1893): 556.

Millican, Albert. *Travels and Adventures of an Orchid Hunter: An Account of Canoe and Camp Life in Colombia, While Collecting Orchids in the Northern Andes*. London: Cassell, 1891.

Mivart, St. George Jackson, and Making of America Project. *On the Genesis of Species by St. George Mivart, F.R.S.* New York: D. Appleton, 1871. doi:10.5962/bhl.title.30543.

Mohan Ram, H. Y. "On the English Edition of Van Rheede's *Hortus Malabaricus* by K. S. Manilal (2003)." *Current Science* 89, no. 10 (2005): 1672–80.

Moskowitz, Samuel, ed. *Science Fiction by Gaslight: A History and Anthology of Science Fiction in the Popular Magazines, 1891–1911*. 2nd ed. Westport, CT: Hyperion, 1974.

Müller-Wille, Staffan. "Collection and Collation: Theory and Practice of Linnaean Botany." *Studies in History and Philosophy of Biological and Biomedical Sciences* 38 (2007): 541–62.

———. "Systems and How Linnaeus Looked at Them in Retrospect." *Annals of Science* 70, no. 3 (2013): 305–17.

Neil, James. *Rays from the Realms of Nature: Or, Parables of Plant Life*. 4th ed. London: Cassell, Petter, Galpin, 1879 (1884).

Nelson, E. Charles. *John Lyons and His Orchid Manual*. Kilkenny, Ireland: Boethius, 1983.

Niles, Grace Greylock. *Bog-Trotting for Orchids*. New York: G. P. Putnam's Sons, 1904.

"[Obituary] Col. M. J. Godfery." *Nature* 155, no. 3943 (1945): 627.

"Orchid Hunting: A Dangerous Pastime." *Singapore Free Press and Mercantile Advertiser*, October 18, 1901, 3.

Orlean, Susan. *The Orchid Thief*. London: Vintage, 2000.

Otten, Charlotte F. "Ophelia's 'Long Purples' or 'Dead Men's Fingers.'" *Shakespeare Quarterly* 30, no. 3 (1979): 397–402.

Paley, W. *Natural Theology: Or, Evidences of the Existence and Attributes of the Deity*. 12th ed. London: J. Faulder, 1809.

Parkinson, John. *Paradisi in Sole Paradisus Terrestris, or, a Garden of All Sorts of Pleasant Flowers Which Our English Ayre Will Permitt to Be Noursed Vp: With a Kitchen Garden of All Manner of Herbes, Rootes, & Fruites, for Meate or Sause Vsed with Vs, and an Orchard of All Sorte of Fruitbearing Trees and Shrubbes Fit for Our Land Together with the Right Orderinge Planting & Preseruing of Them and Their Vses & Vertues*. London: Humfrey Lownes and Robert Young, 1629. doi:http://dx.doi.org/10.5962/bhl.title.7100.

Parkinson, John, J. Mackie, and University of Cambridge Dept. of Plant Sciences. *Theatrum Botanicum: The Theater of Plants: Or, an Herball of a Large Extent; Containing Therein a More Ample and Exact History and Declaration of the Physicall Herbs and Plants That Are in Other Authours, Encreased by the Accesse of Many Hundreds of New, Rare, and Strange Plants from All the Parts of the World, with Sundry Gummes, and Other Physicall Materials, Than Hath Beene Hitherto Published by Any before; and a Most Large Demonstration of Their Natures and Vertues; Shewing Withall the Many Errors, Differences, and Oversights of Sundry Authors That Have Formerly Written of Them; and a Certaine Confidence, or Most Probable Conjecture of the True and Genuine Herbes and Plants. Distributed into Sundry Classes or Tribes, for the More Easie Knowledge of the Many Herbes of One Nature and Property, with the Chiefe Notes of Dr. Lobel, Dr. Bonham, and Others Inserted Therein*. London: Tho. Cotes, 1640.

Parrinder, Patrick. "The Old Man and His Ghost: Grant Allen, H. G. Wells and Popular Anthropology." In *Grant Allen: Literature and Cultural Politics at the Fin de Siècle*, edited by William Greenslade and Terence Rodgers. Aldershot: Ashgate, 2005.

Pliny, the Elder. *Natural History*. Translated by H. Rackham. 10 vols. London: Heinemann, 1968.

Porter, Roy, and Mikuláš Teich, eds. *Drugs and Narcotics in History*. Cambridge: Cambridge University Press, 1995.

Preus, Anthony. "Drugs and Psychic States in Theophrastus' *Historia Plantarum* 9.8–20." In *Theophrastean Studies: On Natural Science, Physics and Metaphysics, Ethics, Religion, and Rhetoric*, edited by William Wall Fortenbaugh and Robert W. Sharples, 76–99. New Brunswick, NJ: Transaction, 1988.

Proust, Marcel. *Swann's Way*. Translated by C. K. Scott Moncrieff and Terence Kilmartin. Vol. 1 of *In Search of Lost Time*. London: Vintage, 1913 (2005).

Rain, Patricia. *Vanilla: The Cultural History of the World's Favorite Flavor and Fragrance*. Los Angeles: Jeremy P. Tarcher, 2004.

Reinheimer, H. *Symbiosis: A Socio-Physiological Study of Evolution*. London: Headley Brothers, 1920.

Reinikka, Merle A. *A History of the Orchid*. 2nd ed. Portland, OR: Timber, 1995.

Rendall, Jane. "The Citizenship of Women and the Reform Act of 1867." In *Defining the Victorian Nation: Class, Race, Gender and the Reform Act of 1867*, edited by Catherine Hall, Keith McClelland. and Jane Rendall, 119–78. Cambridge: Cambridge University Press, 2000.

Richards, Evelleen. "Darwin and the Descent of Women." In *The Wider Domain of Evolutionary Thought*, edited by David Oldroyd and Ian Langham, 57–111. Dordrecht: D. Reidel, 1983.

Rolfe, Robert Allen. "*Cattleya labiata* and Its Habitat." *Orchid Review* 8, no. 96 (1900): 362–65.

[Rolfe, Robert Allen?]. "Dies Orchidianae [Demon Flowers]." *Orchid Review* 4, no. 44 (1896): 233–35.

Russan, Ashmore, and Frederick Boyle. *The Orchid Seekers: A Story of Adventure in Borneo*. London: Frederick Warne, 1897.

———. *The Riders; or, through Forest and Savannah with the Red Cockades*. Edited by Frederick Boyle. London: Frederick Warne, 1896.

Rzepka, Charles J. "'I'm in the Business Too': Gothic Chivalry, Private Eyes, and Proxy Sex and Violence in Chandler's *The Big Sleep*." *Modern Fiction Studies* 46, no. 3 (2000): 695–724.

Sachs, Julius. *History of Botany, 1530–1860*. Translated by Henry E. F. Garnsey. Oxford: Oxford University Press, 1906.

Sanders, Dawn. "Carnivorous Plants: Science and the Literary Imagination." *Planta Carnivora* 32, no. 1 (2010): 30–34.

Sapp, Jan. *Evolution by Association: A History of Symbiosis*. New York: Oxford University Press, 1994.

Schiebinger, Londa. "Gender and Natural History." In *Cultures of Natural History*, edited by Nicholas Jardine, James A. Secord, and Emma Spary, 163–77. Cambridge: Cambridge University Press, 1996.

———. "The Private Life of Plants: Sexual Politics in Carl Linnaeus and Erasmus Darwin." In *Science and Sensibility: Gender and Scientific Enquiry, 1780–1945*, edited by Marina Benjamin, 121–43. Oxford: Blackwell, 1991.

Schneekloth, Lynda H. "Plants: The Ultimate Alien." *Extrapolations* 42, no. 3 (2001): 246–54.

Schweinfurth, Charles. "Classification of Orchids." In *The Orchids: A Scientific Survey*, edited by Carl Leslie Withner, 15–43. New York: Ronald Press, 1959.

Seed, David, ed. *A Companion to Science Fiction*. Oxford: Blackwell, 2005.

Shapin, Steven. *The Scientific Revolution*. Chicago: University of Chicago Press, 1996.

Sloane, Hans. *A Voyage to the Islands Madera, Barbados, Nieves, S. Christophers and Jamaica: With the Natural History of the Herbs and Trees, Four-Footed Beasts, Fishes, Birds, Insects, Reptiles, &C. Of the Last of Those Islands; to Which Is Prefix'd, an Introduction, Wherein Is an Account of the Inhabitants, Air, Waters, Diseases, Trade, &C. Of That Place, with Some Relations Concerning the Neighbouring Continent, and Islands of America. Illustrated with Figures of the Things Described, Which Have Not Been Heretofore Engraved*. London: B. M. for the author, 1707–1725.

Smith, Gerard Edwards. *Catalogue of Rare or Remarkable Phaenogamous Plants, Collected in South Kent*. London: Longman, Rees, Orme Brown, and Green, 1829.

Smith, Jonathan. *Charles Darwin and Victorian Visual Culture*. Cambridge Studies in Nineteenth-Century Literature and Culture. Cambridge: Cambridge University Press, 2009.

———. "Grant Allen, Physiological Aesthetics, and the Dissemination of Darwin's Botany." In *Science Serialized: Representations of the Sciences in Nineteenth-Century Periodicals*, edited by Sally Shuttleworth and Geoffrey N. Cantor, 285–305. Cambridge, MA: MIT Press, 2004.

———. "Une Fleur du Mal? Swinburne's 'the Sundew' and Darwin's Insectivorous Plants." *Victorian Poetry* 41 (2003): 131–50.

Spary, E. C. *Eating the Enlightenment: Food and the Sciences in Paris*. Chicago: University of Chicago Press, 2012.

Sprague, Thomas Archibald. "The Evolution of Botanical Taxonomy from Theophrastus to Linnaeus." In *Linnean Society's Lectures on the Development of Taxonomy*. London: Linnean Society of London, 1948–49.

Stannard, Jerry. "Pliny and Roman Botany." *Isis* 56, no. 4 (1965): 420–25.

Stearn, William T. "Two Thousand Years of Orchidology." In *Proceedings of the Third World Orchid Conference*, 26–42. London: Royal Horticultural Society, 1960.

Steenis, C. G. G. J. van, and M. J. van Steenis-Kruseman. *Flora Malesiana*. Series I, Spermatophyta. Edited by C. G. G. J. van Steenis. Vol. 1. Djakarta: Noordhoff-Kolff, 1950. doi:10.5962/bhl.title.40744.

Stevens, Peter F. *The Development of Biological Systematics: Antoine-Laurent de Jussieu, Nature, and the Natural System*. New York: Columbia University Press, 1994.

Steyermark, Julian A., and Bruce K. Holst. "Flora of the Venezuelan Guayana, VII: Contributions to the Flora of the Cerro Aracamuni, Venezuela." *Annals of the Missouri Botanical Garden* 76, no. 4 (1989): 945–92.

Stiles, W. A. "Orchids." *Scribner's Monthly* 15, no. 2 (1894): 190–203.

Stott, Rebecca. *Fabrication of the Late Victorian Femme Fatale: The Kiss of Death*. Basingstoke: Macmillan, 1992.

Stout, Rex. *The League of Frightened Men*. New York: Bantam, 1935 (1985).

Sutton, Geoffrey V. *Science for a Polite Society*. Boulder, CO: Westview, 1995.

Swinson, Arthur. *Frederick Sander: The Orchid King*. London: Hodder and Stoughton, 1970.

Taylor, John Ellor, ed. *The Sagacity and Morality of Plants: A Sketch of the Life and Conduct of* the Vegetable Kingdom . . . New edition, etc. London: George Routledge and Sons, 1884 (1904).

Tegetmeier, William Bernhard. "Darwin on Orchids." *Weldon's Register of Facts and Occurrences* (1862).

T. F. H. *Feeble Faith, a Story of Orchids*. London: Hodder and Stoughton, 1882.

Thiselton-Dyer, William Turner, and Daniel Oliver. "Report of the Botany of Mr. H. O. Forbes's Expedition to Timor-Laut." *Journal of the Linnean Society (Botany)* 21 (1885): 370–74.

Trapp, J. B. "Dioscorides in Utopia." *Journal of the Warburg and Courtauld Institutes* 65 (2002): 259–61.

Turner, William *A New Herball: Parts II and III*. Vol. 2. Cambridge: Cambridge University Press, 1568 (1992).

Tyler-Whittle, Michael Sidney. *The Plant Hunters, Being an Examination of Collecting, with an Account of the Careers & the Methods of a Number of Those Who Have Searched the World for Wild Plants*. Philadelphia: Chilton, 1970.

Ullery, Jacob G., Redfield Proctor, Charles H. Davenport, Hiram Augustus Huse, and Levi Knight Fuller. *Men of Vermont: An Illustrated Biographical History of Vermonters and Sons of Vermont*. Brattleboro: Transcript Publishing Co., 1894.

Vandermeulen, John H. "Nero Wolfe — Orchidist Extraordinaire." *American Orchid Society Bulletin* 54, no. 2 (1985): 142–49.

Van Dover, J. K., ed. *The Critical Response to Raymond Chandler*. Westport, CT: Greenwood, 1995.

Varey, Simon, and Rafael Chabrán. "Medical Natural History in the Renaissance: The Strange Case of Francisco Hernández." *Huntington Library Quarterly* 57, no. 2 (1994): 124–51.

Veitch, Harry James. "A Retrospect of Orchid Culture." In *A Manual of Orchidaceous Plants: Cultivated under Glass in Great Britain*, 109–27. London: James Veitch and Sons, 1887–94.

Vereecken, Nicholas J., and Ana Francisco. "Ophrys Pollination: From Darwin to the Present Day." In *Darwin's Orchids: Then and Now*, edited by Retha Edens-Meier and Peter Bernhardt, 47–67. Chicago: University of Chicago Press, 2014.

Waddell, Mark A. "Magic and Artifice in the Collection of Athanasius Kircher." *Endeavour* 34, no. 1 (2009): 30–34.

Walters, S. M. "The Name of the Rose: A Review of Ideas on the European Bias in Angiosperm Classification." *New Phytologist*, no. 104 (1986): 527–46.

Wells, H. G., David Y. Hughes, and Harry M. Geduld. *A Critical Edition of the War of the Worlds: H.G. Wells's Scientific Romance*. Bloomington: Indiana University Press, 1993.

Wells, Herbert George. *Experiment in Autobiography. Discoveries and Conclusions of a Very Ordinary Brain (since 1866)*. Philadelphia: J. B. Lippincott, 1934 (1967).

Wharton, Don. "An Orchid a Minute." *Saturday Evening Post* 213, no. 41 (1941): 16–84.

White, Fred M. "The Purple Terror." *Strand Magazine* 18, no. 105 (1896): 242–51.

Wilde, Oscar. *The Picture of Dorian Gray*. Harmondsworth: Penguin, 1891 (1985).

Williams, Benjamin Samuel. *The Orchid-Grower's Manual: Containing Brief Descriptions of Upwards of Four Hundred Species and Varieties of Orchidaceous Plants: Together with Notices of Their Times of Flowering and Most Approved Modes of Treatment: Also, Plain and Practical Instructions Relating to the General Culture of Orchids: And Remarks on the Heat, Moisture, Soil, and Seasons of Growth and Rest Best Suited to the Several Species. Seventh Edition, Enlarged and Revised to the Present Time by Henry Williams*. London: Victoria and Paradise Nurseries, 1894.

Williams, Henry. "[Obituary] Bernard S. Williams." *Orchid Album* 9 (1891).

Wyndham, John. *Day of the Triffids*. Harmondsworth: Penguin, 1951 (1954).

Zirkle, Conway. "The Death of Gaius Plinius Secundus (23–79 A.D.)." *Isis* 58, no. 4 (1967): 553–59.

Index

Page locators given in *italics* refer to figures.